INNOVATIVE EXTENSION APPROACHES IN TECHNOLOGY TRANSFER

Editors

Dr. B.S. Hansra
Dr. P.N. Ananth
Dr. P.T. Suraj
Mr. John Jo Varghese
Dr. S. Prabhukumar
Dr. M.J. Chandragowda

DISCOVERY PUBLISHING HOUSE
NEW DELHI

First Published - 2005

Reprinted - 2018

ISBN: 978-81-8356-018-4

Innovative Extension Approaches in Technology Transfer

Published by:

DISCOVERY PUBLISHING HOUSE PVT. LTD.
4383/4B, Ansari Road, Darya Ganj
New Delhi-110 002 (India)
Phone: +91-11-23279245, 43596064-65
Fax: +91-11-23253475
E-mail: discoverypublishinghouse@gmail.com
sales@discoverypublishinggroup.com
web: www.discoverypublishinggroup.com

Printed at:
Infinity Imaging Systems
Delhi

Dedicated to

The Farming Community

Foreword

It is a great satisfaction for me to write a foreword for a book on *Innovative Extension Approaches in Technology* Transfer, which is the outcome of the National Seminar held at Mitraniketan KVK on February 21-22, 2004 on this subject. The contents of the book and its kind has become imperative as public, private, NGOs and other voluntary action groups face the restriction in amplification of the level of technology transfer to a mammoth target group scattered at various agro-climatic zones due to varied reasons. One such prominent bottleneck can be fingered out "the fiscal discrepancy". With such restraint and the only asset in hand the "efficient manpower," an NGO, Mitraniketan has found out different alternative extension approaches to reach the target groups speckled out in the state of Kerala. I am confident that the R&D institutions across the nation have the similar thrust in the present and future to devise out extension approaches in different ways to reach the end users at any cost. Different types of organizations nowadays churn out the business of technology transfer, though it is highly vested with the public sector. The success of any organization depends on the quality and quantity of technology transferred at the end user level, which is fairly measurable with their impact by their deep percolation into the society. I am in high spirits to note that nation wide participants working on technology transfer, refinement and training attended the National Seminar envisaged by KVK Mitraniketan was really an encouraging sign for the issue. I am confident that this book will throw light on the subject and concurrently will be highly useful to different R&D

workers in technology transfer to understand the problem in depth and devise alternative approaches. I am thankful to my co-workers who have worked sincerely with commitment to organize a meaningful and timely scientific deliberation and most appropriately to bring out the proceedings of the deliberation as a book. I wish readers of the book to cherish with the ideas proposed by different authors. I will fail in my duty if I don't congratulate all the contributors of the articles for this valid and useful book.

Shri. K. Viswanathan
Director,
Mitraniketan & Chairman KVK

Preface and Acknowledgement

The National Seminar on Alternative Extension Approaches in Technology Transfer held at Mitraniketan KVK on 21-22, February 2004 was one of the most successful scientific deliberations in the recent past. Most of the successful alternative extension approaches in technology transfer were presented and highlighted. There was a tremendous response from various R& D institutions to present their approaches for sharing at a common platform without any prejudice. The business of technology transfer is the secret of success to every R&D institutions, but this was a platform envisaged to share experiences. We take the opportunity to thank the entire authors for the contribution of their articles to this book.

The Editorial Board took a challenge to bring out the proceedings of the national seminar and we are highly indebted to all the contributors who made our work to be a perfect pearl. We are thankful to all the staff members of Mitraniketan KVK who have taken severe pains to help us in bringing out this book as a valid document for effective use. The Editors are highly grateful to Discovery Publishing House, New Delhi in accepting to publish this book.

Editors

Contents

Foreword

Preface and Acknowledgements

List of Contributors

1. Innovative Extension Approach of Mitraniketan KVK — 1

   P.T. Suraj and P.N. Ananth

2. Agriculture Information System—Retrospect and Prospects — 20

   J. Vasanthakumar

3. Relevance and Roles of Private Extension in the Context of Millennium Development Plans — 29

   Chandra Gowda, M.J., Prabhukumar S. and N.S. Shivalinge Gowda

4. Cyber Extension: An Alternative Approach in Agricultural Extension-Some Experiences — 49

   Sharma, V.P and Lakshmi Murthy

5. Using ICTs in Development: Information Village Research Project, Pondicherry — 67

   Senthilkumaran, S., Arunachalam, S., Rajamohan, K.G., Rajasekarapandy, R., Gobu, J. Pakkialatchumy, P. Manikandan, M., Rajamanikkam, R., Sivakumar, P. and K. Rameswaran

6. Strategy Focused Knowledge Management Support for Extension Services—The KISSAN-Kerala Approach 86

Srivathsan, K.R., Hariprasad and Ajith Kumar

7. ITC e Choupal 103

Sudipta K. Mukhopadhyay

8. Information and Communication Technology (ICT) as an Alternative Extension Approach: A Critical Analysis of ICT Projects in India 114

Shaik. N. Meera, Anita Jhamtani and DUM Rao

9. Implementation Strategy in World Bank Funded "Diversified Agriculture Support Project (DASP)" in State of Uttar Pradesh with Emphasis on Multiple Approach of Extension for Dissemination of Technology 132

Surya Pratap Singh and Mukesh Gautam

10. Extension and Farmer Training of E.I.D Parry (I) Ltd. 145

Pillai, S.S

11. The Agriclinics and Agribusiness Centres Scheme: An Innovative Extension Approach 155

Alexander George and S. Bhaskaran

12. Role of NABARD in Agriculture and Rural Development 166

Sahoo, B.B

13. Extension Approaches of State Bank of Travancore for Empowerment of Rural Sector 186

Subramoniam, K.

14. Institution Village Linkage Programme—An Alternate Strategy for Technology Assessment and Transfer 198

Ramanathan, S. and Ananatharaman, M.

15. e-Extension on Transfer of Technology—An Experience 215

F.R. Sheriff

16. Participatory Curriculum Development Through Farmer Field School 225

M. Anantharman and S. Ramanathan

List of Contributors

Dr. P.T. Suraj, Training Associate (Animal Science), Mitraniketan Krishi Vigyan Kendra, Vellanad, Thiruvananthapuram , Kerala 695 543.

Dr. P.N. Ananth, Training Organiser, Mitraniketan Krishi Vigyan Kendra, Vellanad, Thiruvananthapuram, Kerala 695 543.

Dr. J. Vasanthakumar, Professor and Head, Department of Agricultural Extension and Rural Sociology, Faculty of Agriculture, Annamalai University, Annamalai Nagar Tamilnadu 608 002.

Dr. M.J. Chandra Gowda, Senior Scientist, Zonal Co-ordinating Unit, ICAR TOT Projects, Zone VIII Bangalore, Karnataka.

Dr. S. Prabhukumar, Zonal Coordinator, Zonal Co-ordinating Unit, ICAR TOT Projects, Zone VIII Bangalore, Karnataka.

Dr. N.S. Shivalinge Gowda, Associate Professor Department of Agricultural Extension, College of Agriculture, GKVK, Bangalore, Karnataka.

Dr. V.P. Sharma, Director (Information Technology, Documentation and Publication) National Institute of Agricultural Extension Management (MANAGE) Rajendranagar, Hyderabad, Andhra Pradesh 500 030.

Mrs. Lakshmi Murthy, Assistant Director (Documentation) National Institute of Agricultural Extension Management (MANAGE) Rajendranagar, Hyderabad, Andhra Pradesh 500 030.

Dr. S. Senthilkumaran et.al., MS Swaminathan Research Foundation, Taramani, Chennai, Tamilnadu.

Dr. K.R. Srivathsan, Director, Indian Institute of Information Technology and Management-Kerala, Technopark, Thiruvananthapuram, Kerala.

Mr. Ajith Kumar, Agricultural Officer, KISSAN-Kerala Project, Indian Institute of Information Technology and Management-Kerala, Technopark, Thiruvananthapuram, Kerala.

Mr. C. Hariprasad, Agricultural Officer, KISSAN-Kerala Project, Indian Institute of Information Technology and Management-Kerala, Technopark, Thiruvananthapuram, Kerala.

Mr. Sudipta K. Mukhopadhyay, ITC Ltd-IBD, Hyderabad, Andhra Pradesh.

Dr. Shaik. N. Meera, Scientist, Division of Extension Communication and Training, Central Rice Research Institute, Cuttack, Orissa.

Dr. Anita Jhamtani, Principal Scientist, Division of Agricultural Extension, Indian Agricultural Research Institute (IARI), New Delhi.

Dr. DUM Rao, Senior Scientist, Division of Agricultural Extension, Indian Agricultural Research Institute (IARI), New Delhi.

Dr. Surya Pratap Singh, Secretary Coordination, Government of Uttar Pradesh & Project Coordinator, Diversified Agriculture Support Project (DASP), Lucknow, Uttar Pradesh.

Mr. Mukesh Gautam, Technical Coordinator, Diversified Agriculture Support Project (DASP), Government of Uttar Pradesh, Lucknow, Uttar Pradesh.

Dr. S.S. Pillai, EID Parry (I), Ltd., Bangalore

Dr. Alexander George, Assistant Professor, Central Training Institute, Kerala Agricultural University, Mannuthy, Thrissur, Kerala 680651.

Dr. S. Bhaskaran, Associate Professor, Central Training Institute, Kerala Agricultural University, Mannuthy, Thrissur, Kerala 680651.

Dr. B.B. Sahoo, Manager, NABARD, Kerala Regional Office, Thiruvananthapuram, Kerala.

Mr. K. Subramoniam, Manager (Training), Staff Training Centre, State Bank of Travancore, Poojappura, Thiruvananthapuram, Kerala 695 012.

Dr. S. Ramanathan, Principal Scientist, Social Science Division, Central Tuber Crops Research Institute, ICAR, Sreekariyam, Thiruvananthapuram, Kerala 695 017.

Dr. M. Anantharaman, Principal Scientist and Head, Social Science Division, Central Tuber Crops Research Institute, ICAR, Sreekariyam, Thiruvananthapuram, Kerala 695 017.

Dr. F.R. Sheriff, Director of Extension, Directorate of Extension Education, Tamilnadu Veterinary and Animal Sciences University, Chennai, Tamilnadu 600 051.

Innovative Extension Approach of Mitraniketan KVK

P.T. Suraj and P.N. Ananth

ABSTRACT

Mitraniketan is a well known Kerala based Non-governmental Organisation. The vision of Mitraniketan is to evolve an alternative development paradigm with a focus on sustainable rural development. The Rural Extension Sub Centre Concept (RESC) of Mitraniketan KVK is accepted as an innovative extension approach succeeded in Indian condition. The rural youths trained in artificial insemination programme of Mitraniketan KVK is facilitated to establish a RESC with a village level organisation. The rural youths trained with a multidisciplinary approach act as an extension worker under the guidance of KVK scientists. This approach not only enables the technicians to earn their livelihood, but also helps the organisation to have access to a large number of farmers. Each RESC in its operational area has contributed much to agricultural development especially animal husbandry activities of the region in various forms like providing timely AI facility at farmers door steps, helping to increase the number of high producing cows and advising the farmers about the scientific management of animals. This paper in detail explains the approach and impact created in the tarsetted area.

Introduction

Mitraniketan is well known in India and abroad as a Kerala—based voluntary Non-Governmental Organisation that has pioneered people-centered holistic rural development for improving the quality of life and living of village communities. It strives to promote rural development with a human face. The founder of the organisation, is the present Director Shri.K.Viswanathan, a Gandhian and a well-known social worker in India and abroad.

The vision of Mitraniketan is to evolve an alternative development paradigm with a focus on sustainable rural development. Mitraniketan has an impressive sprawling campus, which is located on an undulating terrain, with evergreen trees all over. It lies in the lap of nature as it were. The buildings of Mitraniketan, housing various activities are strikingly simple and have been built employing cost-effective construction technology. Their architectural style has a rural character, which blends with the locale. Ecological and environmental idiom is manifested in the campus.

Mitraniketan has a resident community engaged in a variety of activities. It has an unmistakable international presence in a Kerala-setting and promotes a life style that is essentially spartan. Mitraniketan has an activity-pattern that is experimental and innovative as also close knit. It attempts to render thoughts and action proximate and transparent. It's vision and mission address the cause of a sustainable peaceful society.

Mitraniketan is an education-based community, which imparts community—based education through participatory teaching-learning methods and emphasise the dignity of labor. A variety of vocations for rural employment are promoted at Mitraniketan, with efficiency—additions through appropriate science and technology inputs. Mitraniketan also strives to promote and enhance rural human resource development capabilities that can foster a participative culture and work ethics. In its development-update and innovative endeavors, a band of like minded fraternity resident in Mitraniketan and elsewhere lends devoted support.

Mitraniketan is a non-political and secular organisation and is represented in various forums of the Kerala State Government and the Government of India in rural development planning and allied areas. It has had the associations of organisations like OXFAM, Action Aid, EZE of Germany, DST, ICAR, CAPART, HUDCO and the British Council from time to time. The association of Danish Folk High Schools and Danida are actively associating with Mitraniketan in its latest innovative education venture, the Mitraniketan People's college.

Mitraniketan has evoked solid responses from the Germany-based Heisenberg Gymnasium and a few other academic communities from abroad. Many sections of the society from India and abroad visit Mitraniketan for short as well as long duration stay for living experience and to pursue their calling, be it voluntary service or academic studies. They are charmed by the tranquility of the place. The recognitions and awards won by the organisation for the service rendered for the society are numerous and the organisation works with the long-standing experience that it has received over the years.

Mitraniketan has won numerous awards in recognition of its sustained and holistic rural development activity forays, fostering people's participative culture at the grass-roots level, and cultivating a scientific temper. The awards include the prestigious *Jamanlal Bajaj Award, The Indian Merchants Chamber Platinum Jubilee, The KP Goenka Award* and *Rabindranath Puraskar*.

Mitraniketan Krishi Vigyan Kendra (Farm Science Center) is a fully sponsored Technology Transfer project of the Indian Council of Agricultural Research, New Delhi (ICAR) with a mandate of technology transfer, refinement and training in agriculture and other allied fields. Mitraniketan is the pioneer institute to start a KVK in 1979, immediately after five years of the establishment of the first KVK in the country under a Non Government Organisation. At present there are 344 KVKs all over the country with massive infrastructure and qualified manpower. The main mandates of the KVK are technology transfer, refinement and training on agriculture and related sectors. The target groups of the KVK are the practicing farmers, farm women, rural youth and extension functionaries. As per the mandate this KVK has to cater the needs of the various natural resource users who utilize land and water for crop, animal and other associated activities of Thiruvananthapuram district, the capital city of Kerala. The training programmes imparted in this KVK are need based. The frequent surveys and group approach techniques conducted to assess training needs of the farmers and other natural resource users play a major role in the identification and designing of training programmes for the benefit of various target groups of different strata. The technological back up of the institute are the technologies generated by the National Agricultural Research System of the country and the host institute Mitraniketan. The technologies generated under the NARS are being refined and popularized through the on-farm trials and front line demonstrations conducted at the farmer's field.

On an average, 7000 farmers and other natural resource users are being trained every year with the prime goal of economic and social empowerment with 170 courses on varied aspects. The strong manpower under the Animal Science, Horticulture, Home Science, Agricultural Engineering, Agronomy, Agricultural Extension and Fisheries divisions have marked a high appreciation with the adoption level of the technologies at the field level.

The technologies identified and popularized among the farming community are mushroom culture, vermi compost, coir pith compost, sericulture, tissue culture, nursery techniques, post harvest technologies in fruits and vegetables, low cost paddy winnower, advanced production technologies in food and fodder crops, azolla culture, beekeeping, integrated pest management, artificial insemination in animals, fish culture, ornamental fish culture and other associated technologies.

The KVK has full-fledged demonstration units for the above-mentioned technologies for effectual skill based training for the farmers of the district. The KVK is serving as the production input delivery agency for the farming community of the district. The demand for the planting materials, mushroom spawn, quality vermi compost fertilizer packets, coir pith compost, semen for artificial insemination, ornamental fishes etc. falls to be cornerstone of this KVK in helping the farmers of the district.

It is well known that the Rural Extension Sub Center (RESC) concept of Mitraniketan KVK is one of the alternative extension approach as tried and succeeded in India. The rural youth trained in the artificial insemination programme of Mitraniketan KVK, Animal Science Section by the veterinary expert is being facilitated to form a sub center with the village level organisations. The concept is to help the rural youth trained in the artificial insemination towards self-employment. They use this innovative institution to gain employment on all bases. With this innovative approach the rural youth trained from Mitraniketan KVK is able to earn about Rs. 3000 /month as an additional income. At present there are 32 RSECs all along the Trivandrum and adjoining districts, which is expected to increase by 40 shortly.Competent techniques for transferring technologies to target users are a key towards an all-round development of any sector. There are unique methods of transmitting technologies by various organisations and documentation of these approaches has been an untapped area. Sharing experiences in the method of transferring technologies from various Research and Development (R&D) centers is highly needed, as the number of beneficiaries for the target technology is mammoth. A two-step training magnification approach of Mitraniketan

KVK is one of the innovative and alternative extension approaches ever since the inception of this institution for the farming and other allied communities. It is high time that we need to disseminate this approach at various levels and share the experience. It is well known that there are varied extension approaches envisaged by R&D centers across the country, which need to be discussed under a common platform for healthy imitation. Realizing this fact, the country's Best KVK as titled by the Indian Council of Agricultural Research, (ICAR), New Delhi, "Mitraniketan Krishi Vigyan Kendra" attempted to conduct a national seminar on the above said subject for effectual reflection. The participants of the seminar were representatives from the public, private, semi government, NGOs, SHGs, voluntary action groups and other associated agencies related to the development works. The expected output of the two-day national seminar was to draft policy level implications on the alternative extension approachs.

In nutshell, Mitraniketan KVK has piloted in transforming Thiruvananthapuram district of Kerala into a developed agrarian economy. The efforts done to fulfill the mandate of the KVK through an integrated approach have made a distinct positive impact on socio–economic life of rural people. The Indian Council of Agricultural Research acclaimed this fact by presenting the "Best KVK Award" for the biennium 1998-'99 to Mitraniketan KVK for the outstanding contributions in the field of Technology Transfer, Refinement and Training. It is the 25th year of service of Mitraniketan KVK, which has laid roadmap on the development of the farming sector of Thiruvananthapuram district. It has been planned to celebrate the silver jubilee of the institution in 2004 with reference to the service rendered to various sections of the society. One such meaningful celebration was the "National Seminar on Alternative Extension Approaches in Technology Transfer". The proceedings of the seminar is the outcome of this book "Innovative Extension Approaches in Technology Transfer".

Need for Alternative Extension Approaches by KVKs

The need to devise an alternative extension approach for the Krishi Vigyan Kendras have been long felt. The catchment area for providing the services by a KVK is found to be very difficult as it is for a whole district. Mitraniketan KVK is not an exception to that, as it has to cater the needs of natural resource users of four taluks comprising of 84 panchayats, 12 community development blocks with 115 villages. The total population of the district is 29, 38, 706 out of which there are about 1, 01, 599 cultivators and 2, 61, 064 agricultural laborers. The KVK has to satisfy the needs of the above-indicated agrarian population of the district. The manpower of the KVK as sanctioned by ICAR, the sponsoring agency is 16 comprising of six (Training Associates and one Training Organizer) with technical expertise. The ratio of manpower with end users to deliver the needs is very high. These mainly moot the KVKs to devise alternative approaches in designing technology transfer to reach the whole district. It is really a challenge for the KVKs to design methods of their own to create an impact with the mandates of the KVK for a whole district. Many of the KVKs have designed such approaches for their technological reach, which is often not documented. The recent attempts of Prabhukumar and Gowda (2004) from Zonal Coordinating Unit of the TOT projects, ICAR has documented the impact of five KVKs of which Mitraniketan is one amongst them.

Mitraniketan and Alternative Extension Approach

Though a little emphasis has been given in the earlier part of this article on the need for an alternative extension approach and the attempt of Mitraniketan KVK an in-depth analysis is brought in this section. Very few impact studies have been conducted with this innovative approach. Gowda and Samantha (2002) attempted to analyze the impact of this approach with certain biological and social parameters. Rafeekher *et al* (2004) studied the impact of this approach with all possible biological and social parameters to derive the contributions to the district that the KVK has catered with this design. Ananth *et al* (2004) highlighted this innovative approach as an effective tool for performing a holistic approach towards rural development of the district.

The Approach

The innovative extension approach devised by Mitraniketan is the establishment of *Rural Extension Sub Center*. The approach being the establishment of Rural Extension Sub Centers at various parts of the district to have a linkage with the KVK to satisfy the mandates of technology transfer, refinement and training. The sub centres are being manned by the ex trainees of the animal science section who are called as the paratechnicians. Every three months six rural youths are trained in artificial insemination and other management aspects of animal husbandry. The KVK also attempts to train these rural youth occasionally with all other technologies in agriculture with the prime aim of making them empowered to start their own self employed units with the knowledge that they gain. The three-month intensive training with maximum practical sessions make the rural youth self sufficient to perform Artificial Insemination (AI). The KVK distributes certificate to the successful trainees after an examination. Interested rural youths are facilitated with an office at their village to form a Rural Extension Sub Centre (RESC) with the aim of making them to perform AI and distribute the production inputs for agriculture, technical advice and to have link with the KVK. The RESCs perform the job of an agri-clinic to an extent to the farmers of their area. Difficult queries are passed on to the KVK scientists to answer by online. Special projects of the KVK are implemented through these centers, which help the KVK to work at any part of the district. The para technicians are in close touch with the KVK as every monday and thursday of a week they visit here to collect semen for performing the AI.

The *Rural Extension Sub Centers (RSECs)* manned by the ex trainees of the Animal science division of

Mitraniketan KVK have been formed with the following important mandates:

1. Income generation for the para technicians trained in AI
2. Artificial insemination at farmers doorstep
3. Proper follow up after insemination
4. Educating the farmers about feeding practices, care and management of livestock and use of latest technologies
5. To act as a link for conducting farm clinics by KVK Scientists on different aspects of agriculture and associated fields in different parts of the district

A RESC covers an average distance of 15-20 km around the center. It is to be noted that most of them are located in remote villages where no other institutional facility was available in early eighties. However, later in early nineties the public extension system started working out institutions for AI serviced to the animal growers.

Design of the Approach

Liquid semen from the selected pedigree bulls maintained by the animal science section is collected, processed, preserved in the KVK is given to inseminators at 1/10th cost i.e. Rs. 35/10 dose. As indicated above the inseminators regularly collect the semen on every monday and thursday of a week. This facilitates them to meet the demand. A registration book of each cow containing details like name and address of owner, date and number of registration, name of sub center, signature of owner, name and date of birth of cow, age on the day of registration, breed, identification marks, date of purchase, cost of animal, pedigree particulars, lactation particulars, particulars of insemination, calving particulars, vaccinations, result of general examination, treatments for infertility and major ailments, milk recording, de-worming schedule of calves and photo of cow is kept by owner of the cow and the inseminator. Exploiting the higher literacy rate of the state, "owner recording system" of production (milk recording) and reproduction details is being followed. This will be checked by the Training Associate (Animal Science) during his periodical visits. Whenever the animal is sold, this register has to be handed over to the purchaser under intimation to inseminator. After insemination he ensures.

Mitraniketan and Animal Welfare

To be a pioneer in the animal welfare of the district, Mitraniketan has well advanced in the process even before the public extension system in development works. An attempt was taken to elicit the history of the best measures that Mitraniketan has taken to stand as a strong force in the improvement of the cattle development in the district. To improve milk production and to help the farmers in and around the area of Mitraniketan, AI was started in the year 1966. At that time AI was not carried out in this area by the Animal Husbandry Department and this venture took sometime for the farmers to accept. The failure of public extension system in the process has been one of the reasons for the well-received animal husbandry activities of Mitraniketan. Gowda and Samantha(2002) infers that providing service support for livestock management in remote areas, not reached by public extension system in Erode and Trivandrum districts of Tamilnadu and Kerala states has been achieved through Para Technicians trained by KVK MYRADA and KVK Mitraniketan (NGO KVKs) in these districts respectively.

A prologue of activities under taken by Mitraniketan (since 1960) in animal husbandry sector is given below:

1. Used superior Indian breeds of bulls initially and initiated artificial insemination under the supervision of a veterinary doctor.
2. Imported Jersey bull from England in 1965 with the help of Ox-fam Foundation for grading up of the local

cattle all follow up actions based on the details in registration book and educate the farmers about care and management of cows.

3. Started extensive crossbreeding in Thiruvananthapuram district to enhance milk production. Attempts have also been taken in order to popularize cross breeding through AI.

4. Took initiative in establishing cooperative milk societies to assure market and optimum returns.

5. Introduced Holstein Friesian bulls for further improvement of germplasm.

6. Trained rural youths in artificial insemination and animal management.

7. Established Mitraniketan sub centers at different villages to provide artificial insemination facility at farmer's doorsteps.

8. In 1979 Government of India sanctioned Krishi Vigyan Kendra (KVK) under Indian Council of Agricultural Research (ICAR). Since 1979 cattle development activities are coordinated by the Animal Science Section of this KVK under a veterinary doctor with postgraduate degree.

9. Former sub-centers were converted into Rural Extension Sub-Centers (RESC) and at present 35 sub-centers are scattered in Thiruvananthapuram, Kollam and Alappuzha districts of Kerala and the neighboring district Kanyakumari of Tamilnadu state.

10. Veterinary expert of this KVK conducts infertility camps; vaccination camps, and has interaction with farmers and visit farmers field.

11. At present the sub-centers are functioning as an extension agency of the KVK to organize agriclinics, distribution of seeds, planting materials and informing various developmental programmes.

The para technicians, with the help of veterinary expert of KVK provide the information on nutrient and disease management. Various camps on livestock health, vaccination, deworming, infertility and its treatment are organized at appropriate times. As a result, most of the diseases and problems related to parasites and nutritional deficiencies have been brought down. This was accomplished by mobilizing communities on a large scale to get desired results.

Currently, 35 rural youths are working as techno-agents in these RESCs. With an initial investment of less than Rs. 5,000/- on critical tools and equipment and recurring expenditure on semen and the cost of mobility, these para technicians are able to earn a net income of about Rs. 16,500/- a year. This income could be improved further to about Rs. 19,500/- if they could use frozen semen supplied by the Kerala Livestock Development Board, if made regularly available.

The Impact

(a) Cattle Population in the District

It is observed that there is 35% decrease in total cattle population in the last twenty years owing drastic reduction in number of males and reduction in number of nondescript females. However, there is an increase in the percentage of breedable cows which clearly indicate the positive improvement in the area of management of cows and increased facilities available for AI. It is also due to the qualitative improvement in the genetic make up and quantitative reduction in the low performing animals.

(b) Changes in AI over the Years in the Project Area

The number of sub centers and number of AI done during the years is presented in Table 1.1. The increase in number of artificial inseminations performed can be attributed to the increase in number of sub centers of the KVK as well as number of inseminations per inseminators in the area. The data over the years infers that there were

Table 1.1. Number of Sub-centres and Number of AI Done

Year	*No. of Centres*	*No.of AI*
1980-81	10	5498
1981-82	11	3422
1982-83	12	2148
1983-84	13	3025
1984-85	13	4835
1985-86	13	4104
1986-87	13	2855
1987-88	13	3100
1988-89	13	2932
1989-90	13	3161
1990-91	13	2625
1991-92	13	3197
1992-93	13	3017
1993-94	13	5482
1994-95	15	5800
1995-96	15	6200
1996-97	17	6300
1997-98	17	6450
1998-99	17	6600
1999-00	19	6750
2000-01	20	6850
2001-02	22	6972
2002-03	25	7156
2003-04	32	7243

earlier hiccups in establishing the rural sub centers. Since 1985-86 there was a steady growth upto 91-92 in the number of sub centers established by the KVK. Then a constant increase was observed up to 2003-04.

(c) *Changes in the Number of Breedable Cows*

The percentage of breedable cows has increased from 54% to 68% is the impact of the project area that Mitraniketan KVK operates says Gowda and Samantha (2002). An improved data collected by the team that conducted the impact study refers that in 1980 the number of breedable cows were 50% and increased to 80% in 2000. The change in the number of breedable cows is attributed towards the intervention of the KVK with special emphasis on the AI with the quality of semen used by the para technicians of the project area. About 35 rural youths are performing for the betterment in the animal care and development of the district. This has been a direct impact over the changes in the number of breedable cows in the district.

(d) *Age at First Conception (months)*

Age at conception was reduced to 20 months from 28 due to genetic improvement along with intensive awareness programmes in feeding and management, especially due to calf and heifer management. Gowda and Samantha (2002) also infers that the age at the first conception has been reduced as indicated above. The reasons behind such a drastic change have been attributed by owing to regular contacts and technical support during the period.

(e) *Conception Rate*

The conception rate obtained through AI in different parts of India is less than 40%. Though the conception rate during the project period is fluctuating, it never went below 40%. Mitraniketan KVK maintains pedigree bulls with higher semen quality for ensuring higher conception rate. The average conception rate obtained during the project period is 62%.

KVK Intervention in Raising the Conception Rate of Cattle in the District

On the establishment of Krishi Vigyan Kendra, animal husbandry activities of Mitraniketan were taken up under its fold and were continued. The ICAR also approved the training programme of unemployed educated youths as a self-employment measure and the training programme already established were continued. In due course AI sub centers known as Rural Extension Sub Centers were established at various places by para technicians agents. The most important attribute noted was low conception rate in the district. After the intervention of the KVK with the para technicians the conception rate has increased to a greater extent. The reasons for low conception rate in the district can be attributed to the following reasons.

1. *Access to the Services*

Access to the public veterinary extension system in relation to reach the farmer's needs was the main constraint. Animal welfare and development activities in most of the areas in Kerala are largely provided only by the public extension system. It was impossible for farmers who resided even 5 km away from the center to bring their animal to AI center for insemination. Inadequate access for AI purposes is one of the prime reasons for low conception rate of cattle in the district. Lack of institutional facilities and the amenities like vehicles have constrained the government inseminators to do improvement work in often remote villages. Access to the tribal areas in particular is also a handicap for the public extension system to cater the needs.

2. *Timeliness*

Attending the cattle at right time at the right place has a direct impact over increasing the conception rate. It could be observed that the time of insemination was never attended properly due to the inability of farmer to bring the animal to the center. The system of working with a prescribed time schedule by public AI centres and casual attitude of government inseminators results in low conception rate. No alternate arrangement for insemination on government holidays by the public extension system has caused several difficulties.

3. *Poor Quality Semen*

Quality semen is the critical production input while aiming at cross breeding. The careless management of semen in preservation and transport has limited the conception rate. Bulls maintained at several places are not often replaced at the right time. It could be inferred from service providers that the availability of bulls is also on high demand.

4. *Awareness*

Care and management of cattle after insemination are essential for successful results in the conception of animals. The scientific know how to increase the conception is highly needed to be imparted to the animal growers. The service providers in the target area normally neglects passing on the know-how on the care and maintenance of animals after the post insemination phase. The awareness campaigns are veterinary clinic types rather creating awareness on certain critical knowledge items to the animal growers.

These are the reasons for the low conception rate prevailing in the district. Due to the intervention of the KVK this has been enhanced. The para technicians provide the services to the farmers at the door steps without considering the time, reaching the farmer at any time, providing quality semen, advising and imparting the technical know how of the farmers in the after care in the post insemination phase.

Certain cases have been depicted for the better understanding of the rural youth who use the RESC for their livelihood.

CASE 1

Shri Shaji after completing his higher secondary education decided to become a full time dairy farmer, following his father's footsteps. He joined the vocational training course—dairy management and artificial insemination in Mitraniketan Krishi Vigyan Kendra in 2001.

After the completion of the course he started a Rural Extension Sub Centre of Mitraniketan KVK at Balaramapuram. Along with that he expanded his dairy unit into a 10 milch cow unit. As the milk production increased he started a tea shop. As the business flourished, he added five buffaloes and five milch cows to his dairy unit.

His entrepreneurial skills motivated him to start two more hotels. Now he is producing 150 litres of milk per day. The income from the sale of milk alone comes to Rs.1500/ day. His three hotels are having good business with an average turn over of Rs.1,80,000/-month. He is providing employment to 17 males and 3 females.

His Rural Extension Sub Centre was one of the most active sub centers of Mitraniketan KVK. He was very keen to provide the artificial insemination facilities at farmer's doorsteps at Balaramapuram. Because of his busy working schedule and inability to find time to provide round the clock service to the farmers, he sponsored a candidate for the training on dairy management and artificial insemination. Now the recently trained technician is running the Rural Extension Sub Centre at Balaramapuram. Through his dedicated working nature and entrepreneurial skills Shri Shaji has become a model for the youths in his village.

CASE 2

Kottoor is a Tribal Village in Western Ghat region of Thiruvananthapuram district. Nearly one thousand tribal families are living in the interior forest of Kottoor. Though many of the organisations has taken initiative in promoting agricultural activities in the region, most of them failed miserably due to lack of continuous monitoring. In addition to this the insurance coverage given to cattle by a general insurance company was withdrawn pointing out inadequate breeding and treatment facilities available. Usually those who are willing to help the tribals have to walk five kilometers through the dense forest sometimes risking their lives, due to frequent elephant attacks.

Identifying this problem Mitraniketan KVK trained Mr. Sureshkumaran Nair, in artificial insemination in the year 2000. After completing the training Sureshkumaran Nair started a Rural Extension Sub Centre at Kottoor in collaboration with a Youth Club. He has shown the courage to work through the forest for delivering artificial insemination services and first aid to animals kept by the tribals. In addition to this he is regularly supplying mushroom spawn, seeds, planting materials and vermicompost to the farmers of the locality. In collaboration with the youth club he regularly arranges training programmers and group discussions for the farmers. KVK Scientists regularly visits the sub center and interact with the farmers. Through rendering the services Mr. Sureshkumaran Nair earns Rs. 3,000/-month.

Mr. Sureshkumaran Nair has started a mini dairy unit with three cows in his homestead, which provides additional income to his family. He also maintains a medicinal plant nursery with 100 types of medicinal plants.

Conclusion

Application of innovative extension approaches are essential for technology transfer in agriculture and allied fields. Relatively few number of extension workers available in the area makes it essential to devise innovative approaches for providing extension services to a large number of people. Mitraniketan KVK has developed its method of establishing RESCs to increase its reach to the farmers. This approach of strengthening, training and empowerment of rural youths for taking up the responsibility of agricultural and social development has made its impact in the project area in the form of higher breedable cattle population, higher conception rate and increased awareness in management of cows.

References

Chandre Gowda, M.J and Samantha, R.K. 2002. Para technicians as alternative veterinary service for livestock management. LEISA India. 4 (1).

Prabhukumar, S and Chandre Gowda, M.J. 2004. Innovative extension approaches: KVK experiences. ICAR Transfer of Technology Projects, Zonal Coordinating Unit, Bangalore.

Rafeekher, M., Suraj, P.T., Girija, P and Ananth, P.N. 2004. Techno agents in dairy development: Experiences of KVK Thiruvananthapuram. In: Innovative Extension Approaches: KVK Experiences, ICAR Transfer of Technology Projects Zonal Coordinating Unit, Bangalore.

Ananth, P.N., Suraj, P.T and John Jo Varghese. 2004. Rural Extension Sub Centre concept of Mitraniketan Krishi Vigyan Kendra. Presented at the National seminar on frontiers of extension research, 10-11 March, 2004. Department of extension, Annamalai University, India.

2

Agriculture Information System—Retrospect and Prospects

J. Vasanthakumar

ABSTRACT

A review of studies on knowledge dissemination and utilisation revealed three basic models with emphasis on dissemination, problem solving and social interaction and a synthesis of these resulted in 'linkage model'. A system has to be interactive and holistic rather than a set of discreet, parallel or competing institutions (Swanson and Claar, 1983). Agricultural Information System has been proposed as an alternative policy model (Roling, 1988). The Agriculture Information System has to provide consumer oriented information not only to take advantage of farmers' willingness to pay for extension but also to ensure the livelihoods of poor rural population in view of globalisation. The system should facilitate interventions to organise resource poor farmers because contract farming with farmers organisations/association instead of with individual farmers, is likely to benefit all those concerned. The ATMA model could solve some of the problems in transfer of technology that are to be taken care of in any future endeavor. Further, efforts are to be made to eliminate constraints like lack of adequate linkages and absence of feedback mechanism. As the farmers are faced with increasingly complicated decisions, the management decision support system might be tried together with networking of such farmers or their organisations decision support system might be tried together with networking of such farmers or their organisations.

Introduction

Agricultural Extension is an essential mechanism for delivering information as an "input" into modern farming. In order to understand the importance of this input in handling agricultural technologies, systems approach is found useful. The term 'system' refers to an arrangement of components or elements which interact to achieve some common purpose . The term agricultural information system referred to 'a system in which agricultural information is generated, transformed, transferred, consolidated, received and fedback in such a manner that these processes function synergically to underpin knowledge utilisation by agricultural producers' (Roling, 1988). The emphasis is on the institutions, information flow and linkage mechanisms.

Agricultural Information System in India

Agricultural Information System in India may be described under three broad categories, namely, Public Extension Services, Private Extension Services and Mass Media and Information Technology (Government of India, 2000).

The Public Extension Services may include

- State Government operated extension (Departments of Agriculture, Horticulture and Livestock Development, Fisheries, etc.)
- State Agriculture University based extension (Directorates of Extension and Krishi Vigyan Kendras (KVKs), and
- ICAR extension (Zonal Research Stations/Krishi Vigyan Kendras, Agriculture Technology Information

Centres (ATICs), Institute Village Linkage Programme (IVLP), etc.)

The Private Extension Services may include:

- Community Based Organisation (Farmers' Organisations, Farmers' Cooperatives, Self-Help Groups, Farmer Interest Groups, etc.)
- Para Extension Workers (Contact Farmers, Link Farmers, Master Farmers, Mitra Kisans, Mahila Mitra Kisans, etc.)
- Agri Clinics and Agribusiness Centres
- Input suppliers/dealers (Pesticides, Seeds, Nutrients, Farm Implements, etc.)
- Corporate Sector (Commercial crops-tobacco, tea, coffee, oilseeds (sunflower), vegetables, seeds, farm implements—tractors, threshers, sprinklers, drip irrigation, etc.) and
- Non-Governmental Oganisations (NGOs)

Mass Media and Information Technology may include:

- Print Media including Vernacular Press
- Radio, Television (Doordarshan and Private Channels) etc.
- Electronic Connectivity through computers, NICNET, Internet, V-SAT etc.
- Private Portals
- Farm Information and Advisory Centres (FIACs) and
- Public and Private Information Shops

These categories present varied institutional mechanisms for the generation and flow of information to benefit of the farmers ultimately.

Retrospective View of Agricultural Information System

The organisation of Extension Services and their

contact with farmers were often recorded as some what haphazard. They were organized either by central or local Governments or by agricultural colleges in association with experiment stations or by farmers organisations. The extension services were relatively small in scale and limited in the scope of their work. Later, Government funding became more important, their objectives became broader, and the extension workers became better trained and more professional. In addition, agriculture-related commercial companies and agricultural commodity marketing boards entered the scene. They were concerned with the supply and quality of their specific product (Jones and Garforth, 1997). A variety of non-governmental organisations involved in agricultural and rural development came into existence with remarkable performance. The constraints to technology transfer are increasingly recognised as the lack of adequate linkages and absence of feedback mechanisms. It should lead to evolution of a technology development and transfer system through an interactive, holistic system rather than a set of discreet, parallel or competing institutions (Swanson and Claar,1983).

Prospects of Agricultural Information System

In recent times, the focus of information system has often become more diversified. In less developed countries, the main focus remains on food grains production. But there has been a growing recognition of the need to reach and influence small, resource poor farmers. For the commercial farmers of the developed countries, surplus production has become a major problem because the farmers face economic and policy pressures to restrict production. Associated with intensive production methods, many issues and problems regarding sustainability have also arisen.

The rural population will be better educated their exposure to the mass media will continue to reduce their isolation from information and an awareness of their situation with in a national and international context. This exposure will not reduce the need for extension. Rather, given the changing demands on agricultural producers from population growth, increasing urbanisation, globalisation and market requirements, the more knowledgeable farming population will require different kinds cf extension services. It will therefore necessitate more highly trained, specialized, and technically competent workers, who also know where to obtain relevant information (Hayward, 1990; Moris, 1991). The need for agricultural and rural information services is likely to intensify in the foreseeable future.

A recognition of the locale specific nature of farming systems and the agricultural information systems which support them is an important source of the pressure towards devolution of extension services. This recognition also implies that extension workers and farmers be jointly involved in the verification and adaptation of new technology, and thus that the extension workers respect farmers as experimenters, developers, and adapters of technology and devote more energy on communication within their local areas. The devolution of extension services to become local organisations is a reasonable corollary of this developments in mass media technology (Garforth, 1986) which will continue to support this localisation of extension effort.

Moreover, a rapid increase can be expected in the use of information technology in support of extension (Rivera and Gustafson, 1991). The continuing rapid development of telecommunications and computer based infomation technology is probably the biggest factor for change in extension that may facilitate and reinforce other changes. There are many possibilities for the potential applications of the Information Technology in agricultural extension (Zijp, 1994). Information Technology will bring new information services to rural areas over which farmers, as users, will have greater control than over current information channels. Even if every farmer does not have a computer terminal, these could become readily available at local information resource centres, to help farmers to make decisions. However, it will not make extension workers redundant. Rather, they will be able to concentrate on tasks and services where human interaction is essential in helping farmers individually and in small groups to diagnose problems to interpret data, and to apply their meaning (Leeuwis, 1993).

For many years agricultural extension was considered the monopoly of the public sector. However, with the wide range of demands for agricultural technology in the changing scenario there is growing recognition that public extension by itself cannot meet the specific needs of various regions and different classes of farmers. The ideal would be an addition of competitive private market for agricultural information, affordable to all and responsive to farmers' needs. A significant deterrent to expansion of private sector involvement in technology transfer is the provision of subsidised agricultural inputs and services by public agencies. This often leads to the creation of an uneven playing field and discourages market entry by private providers. Wherever possible such subsidies will be phased out in order to stimulate emergence of a private input supply network to provide various inputs and services on a full cost recovery basis.The future extension regime would recognise the role of multiple agencies of different strengths in transfer of technology. Policy environment will promote private extension to operate in roles that complement, supplement, work in partnership and even substitute for public extension. Public responsibility for financing extension on themes such as environmental protection will likely remain, although private delivery modalities may prove useful (Chapman and Trip, 2003).

There is also a need for a more farmer participatory approach in situation analysis, problem diagnosis, search for appropriate technology, designing the process of implementation, monitoring and evaluation, and feedback. The extension agent is no longer seen as the only expert who has all the useful information and technical solutions; and solution to local problems are to be developed in partnership between the extension agent and farmers. Extension workers therefore need new skills and competencies.

Research and extension organisations have not focused on farmers' problems mainly due to lack of effective feedback. The vast majority of small and marginal farmers in India, especially women, lack an effective voice in influencing research and extension priorities. A demand driven extension system has to be created, by providing farmers with access to linkage mechanisms for obtaining relevant information to help them articulate their problems and needs with reference to their production and marketing plans. A key factor in improving feedback is to organize farmers into functional groups. Linkage mechanisms would also ensure meaningful farmer representation in the governing bodies of public and private extension services so as to exert influence on decisions on the planning, implementation and monitoring of development programmes.

Marketing extension has not received the attention it deserves. This assumes greater significance in the light of globalisation under the WTO which opened up export opportunities. Public extension functionaries are presently ill equipped to deal with marketing extension. Private efforts deprive the farmers of their rightful share in the profit. The marketing extension service will need to address the issue of effective marketing through extensive use of media and information technology (Surinder and Vasanthakumar, 2002).

Conclusion

Changes in the socio-economic profile of farmers, changes due to population growth, globalisation and market requirements and changes in national and international policy with emphasis on sustainability placed heavy demands on information systems. Departure from dependency on traditional public extension system to encourage private sector and NGOs for a complementary role envisage bright prospects for information systems. A shift from production oriented information system to production, value addition and marketing oriented information system is bound to occur. Further, taking adavantage of Information Technology, initiatives for cost recovery of extension services, employing participatory approaches in technology generation and transfer indicate that quality of information system will dominate the global scenario in future.

REFERENCES

Government of India (2000). Policy Framework for Agricultural Extension (draft) Ministry of Agriculture, Department of Agricultural and Co-operation, New Delhi.

Chapman, R and R. Trip (2003). Changing Incentives for Agricultural Extension—A Review of Privatised Extension in Practice. Agren Network paper No.132, London Overseas Development Institute.

Garforth, C. (1986). Mass media and communications technology. In G.E. Jones (Ed.) Inversting in rural extension: Strategies and goals London and New York: Elsevier Applied Science Publishers.

Hayward, J. (1990). Agricultural extension: The World Bank's experience and approaches. In FAO report of the global consultation on agricultural extension Rome: FAO.

Jones G.E and C. Garforth (1997). The History, Development, and Future of Agricultural Extension. In: Swanson B.E, R.P. Bentz and A.J. Sofanko (Ed.). Improving Agricultural Extension, Rome, Food and Agriculture Organisation of the United Nations.

Leeuwis, C. (1993). Of computers, myths and modeling: The social construction of diversity, knowledge, information and communication technologies in Dutch horticulture and agricultural extension. Wageningen Studies in Sociology, 36. Wageningen: Agricultural University.

Moris, J. (1991). Extension alternatives in tropical Africa. London: overseas Development Institute.

Roling N.G. (1988). Extension Science—Information Systems in Agricultural Development. Cambridge, Cambridge University Press.

Rivera, W.M., and Gustafson, D.J. (Eds.). (1991). Agricultural extension: Worldwide institutional evolution and forces for change. Amsterdam and New York: Elsevier.

Swanson, B.E. and J.B. Claar (1983). Technology development, Transfer and feedback systems in agriculture: An operational systtems analysis. Proposal for International Programme for Agricultural Knowledge Systems. Urbana (III): University of Illinois, unpublished paper.

Surinder,V. and J. Vasanthakumar .(2002) "Information sources for Market Oriented Agriculture and Horticulture", paper presented in National Seminar on Market Oriented Agriculture,February 7- 8, 2002, Department of Agricultural Extension, Annamalai University,Annamalainagar (mimeo)

Zijp. W. (1994). Improving the transfer and use of agricultural information: A guide to information technology. World Bank Discussion Paper 247. Washington, DC: The World Bank.

3

Relevance and Roles of Private Extension in the Context of Millennium Development Plans

Chandra Gowda, M.J., Prabhukumar S.
N.S. Shivalinge Gowda

ABSTRACT

The UN Millennium Declaration (2000) adopted by 189 countries suggests collective commitments to overcome poverty that still grips most of the world's population. Millennium development goals place human well-being and poverty reduction at the centre of global development. The purpose of development is to improve people's lives by expanding their choices, freedom and dignity. Three other goals correspond to creating essential conditions for human development namely environmental sustainability, promoting gender equity and empowerment of women and enabling global economic environment. Agricultural Extension as a profession has great roles to play in achieving these development goals at different levels within a nation. Agricultural Extension, by definition, is a professional communication intervention deployed by institutions to induce change in voluntary behaviour with presumed public or collective utility. It aims at change in people than change in practices alone. This capacity development through learning, which is more a voluntary process, requires genuine commitment and interest. Perceived poor performance of past investments in extension has forced re-examination of public extension services for its inadequacies in the more competitive, market-oriented climate of today's agriculture and rural development. Decentralisation, privatisation, cost recovery and participation by stake holders within a pluralistic financing and delivery system are some of the major reforms being pursued in extension's current transition. Private extension is seen as a mechanism to reduce economic burden of governments, increase efficiency of extension services and increase accountability of extension system. How the private extension, as part of pluralistic approach, can overcome the existing limitations of the public extension system is a big challenge to be answered for its sustainable existence. Also, desired nature and extent of private extension amidst deluding attempts to claim the status of an alternative to the prevalent system need to be thoroughly analysed. The present paper is an attempt in this direction.

The Background

The status of agricultural extension services could be described as largely in the public sector, operated in an interpersonal mode of select contact farmers and with low level of involvement of farmers in technology development and dissemination process. As it is substantially top-down, there is little scope for localised planning and action. Thus there are several system constraints, which have reduced the extension efficiency (ICAR, 1999).

System Constraints

1. **Multiplicity of Technology Transfer Systems.** A situation exists where each line department operates its own parallel technology transfer system. This duplicative structure of the current system is inefficient.

2. **Narrow Focus of Agricultural Extension System.** Major extension resources have been allocated to mainly cereal crops and few extension resources have been

allocated to other commodities such as oilseeds, pulses, sorghum, millets and horticultural crops, and most livestock enterprises.

3. **Lack of farmer Focus and Feedback**. With the exception of selected NGO operated programmes and dairy farmers who participated in Operation Flood programmes, the vast majority of India's small and marginal farmers, especially women, lack an effective voice in influencing research and extension priorities.

4. **Inadequate Technical Capacity within the Extension Systems**. According to a study by the Ministry of Agriculture with the help of the World Bank, with a few brilliant exceptions, the village extension workers are neither an educated nor a knowledgeable lot and some of them are even illiterates (Shaji, 2001).

5. **Weak Research—Extension Linkages.** Lack of competent subject matter specialists is responsible for the poor research—extension linkages and the lack of integration across crop and livestock systems. These constraints severely limit technology dissemination system in assisting farmers in exploring improved production practices and incorporating high value commodities into their farming systems.

6. **Poor Communication Capacity.** Most technical staff within the line departments lacks the capacity to effectively communicate with both, the research system and the stakeholder groups. There is little use of communication technology.

7. **Inadequate Operating Resources and financial Sustainability**. In the prevailing situation, nearly 90 per cent of extensions recurrent budget is allocated to salaries and personal emoluments. Most line departments do not have sufficient operating funds to carry out routine extension activities nor the resources to maintain existing physical facilities and

equipment, what to talk of upgrading their human resources.

In addition to the above, the line departments (like Department of Agriculture, Department of Horticulture, Department of Animal Husbandry etc.) have the problem of (Rasheed, 2003);

1. Strong linear hierarchy and thinly spread out in area coverage due to relatively few staff at operational level
2. Implementation of a large number of schemes with little time for assisting farmers with advice on solving specific field problem
3. Severe restrictions on mobility due to limited operational contingencies.

The Human Development Report 2003 (UNDP, 2003), which is devoted to the theme of Millennium Development Goals summed up the limitations in the existing extension system as—Agricultural services, if available, mainly come from private firms selling inputs and offering advice that is often incorrect and almost always incomplete. Government agricultural extension services have focused on distributing seeds and fertilizers, often promoting varieties and formulations unsuited to local conditions.

The Arguement for Pluralistic Extension System

Agricultural Extension, by definition, is a professional communication intervention deployed by institutions to induce change in voluntary behaviour with presumed public or collective utility (Roling, 1988).

In India, the extension system has been undergoing a metamorphosis with experiments and innovations in alternate extension approaches. These are gaining ground due to the opportunities available at micro and macro level. At micro level, farmers in their attempt to get benefited from the emerging liberalisation and privatisation situation or to fight out the ill effects of liberalisation and privatisation are looking for multiple sources of information and innovations. At macro level, the financial support to public extension is declining at a rapid rate, thus reducing the activities and reach of public extension, thereby provide space and opportunities for the private extension service providers to play their role.

The workshop on Extension and Rural Development—Converging views on Institutional Approaches organized by the World Bank in 2002, enlisted the following reasons on which the private extension approaches are gaining ground (World Bank 2002) .

The argument for pluralistic extension systems is based on the premise that the private sector (whether private companies, NGOs, RPOs or specialized consulting firms) can provide extension services more efficiently and effectively than public sector agencies, and that these advantages increase the likelihood of long-term and sustainable services.

Private extension is becoming increasingly important because the public sector is withdrawing from some service provision and states are privatizing area deemed to be private goods.

NGOs are often flexible, committed to working with the poor and disadvantaged, able to provide intensive and integrated assistance to target grass-roots, community organisations, and adapt approaches to local situations. They often have skills in building local organisations and linking them to markets.

Rural Producer Organisations (RPOs) empower farmers to express demands, contract service providers who meet their needs, and enhance accountability. It makes sense for an RPO to engage in extension delivery if the RPO strategy is to improve the agricultural productivity of its members, if services have a clear commodity focus, if farming is viewed as a business, and the RPO has the human and financial resources to do so.

The Alternatives

Rural poverty in India is a complex phenomenon and there cannot be one dominant approach for its alleviation (Bhatnagar and Schware, 2000). Many experiments have succeeded in alleviating poverty in cluster of villages through participatory planning, effective supply of credit to the poorest of the poor, improved management of government-run poverty alleviation programmes, and work of some NGOs in building network of self-help groups.

The current scenario is throwing up lots of alternatives. Many governments and organisations have been striving towards innovations in agricultural extension. Almost all of them envisage one or more of the following.

- Participatory extension approaches
- Information and Communication Technology based extension approach
- Private Extension Service approach
- Extension through Farmers Organisations (Rural Producer Organisations)
- Contract extension approach
- Decentralized extension approach
- Community-based Group extension approach

Private extension is a process of funding and delivering the extension services by private individuals or organisations. As many as 15 different players have been identified as possible private extension service providers under Indian situations. Private extension services are seen as supportive to public extension and not as a substitute. Currently, private extension service providers could be categorised as agribusiness companies, farmers' organisations, mass media, Non-government organisations and agricultural consultancy services (Chandrashekara, 2001).

The Millennium Development Goals and their Relevance to Development of India

In 2000, the UN Millennium Declaration, adopted at the largest ever gathering of heads of state, committed countries—rich and poor—to do all they can to eradicate poverty, promote human dignity and equality and achieve peace, democracy and environmental sustainability. The Millennium Development goals call for an attack on inadequate income, widespread hunger, gender inequality and environmental deterioration. The Indian Agricultural Extension, as a profession and as a science has got a major role to play in achieving at least 4 goals (Goals 1,3,7 & 8 specified in HDR Report 2003) as detailed below.

Goal 1: Eradicate Extreme Poverty and Hunger

As per the data available in 2003 HDR report (UNDP, 2003), 34.7 per cent of the Indian population still lives below poverty level, and 24 per cent of the population is undernourished. Though the proportion of rural population out of the total population is declining (from 78.7 per cent in 1975 to 72.1 per cent in 2001), it is a fact that nearly 68 per cent of the Indian population remains in rural areas even after a decade from now.

Addressing poverty issues among others, include expanding poor people's access to land, credit, skills and other economic assets; and increasing small farmers' productivity and diversification. Both the above issues demand a definite and purposeful set of activities from the agricultural extension field.

As many hungry people are landless, or lack secured tenure, they need to be provided with alternate income earning avenues. Low agricultural productivity needs to be addressed, particularly in marginal ecological regions with poor soils and high climatic variability.

Meeting the Millennium Declaration goal for hunger require improving food distribution and increasing production. Top priorities for increasing production include

- Focusing on technologies that raise agricultural productivity. Doing so will also raise incomes for people few assets other than land.
- Directing more resources to agriculture. Poor countries have neglected agriculture—a trend that must be reversed.
- Preventing environmental degradation. New policies and technologies to raise productivity must also protect critical ecosystems. Poor people suffer the most from environmental degradation, but poverty also leads to environmental degradation. In developing countries low productivity is more often the cause of such degradation—while in Europe and North America high productivity is the cause.
- *Sharing resources more equitably*. Women, who produce most of the food consumed in Sub-Saharan Africa and Asia, must have more secure access to land. The same goes for landless people.

Most of the investment in agriculture is on more favored agricultural areas and much less is invested on marginal lands.

As gains of the green revolution have bypassed resource poor situations, a doubly green revolution is needed–one that increases productivity and improves environmental sustainability. Increased investments are needed to research and develop better technologies and disseminate them through extension services.

Goal 3: Promote Gender Equality and Empower Women

The current focus on the development need for social justice and involving people in decisions that affect them and their communities. Promoting gender equality and empowering women, valuable in themselves, because women are agents of development. Better-educated, healthier women contribute to higher productivity for example, by adopting farming innovations.

Goal 7: Ensure Environmental Sustainability

About 57.1 per cent of the geographical area of the country is subjected to one or the other types of degradation, out of which 45.3 per cent is hit by water erosion alone.

India remains in a very dismal and disadvantageous position when it comes to protection of natural system. For example, only 3.4 per cent of people in developed nations are depending on traditional fuel, whereas, it is as high as 20.7 per cent in India.

Ensuing environmental sustainability requires managing ecosystems so that they can provide services that sustain human liveliness. Promoting environmental sustainability depends on the extent to which local people are involved in working out solutions. Resources to invest in agriculture and small-scale manufacturing to improve worker productivity need to be mobilized both at users-group level and at community level.

A community can get more food by converting forest to farmland, but in doing so it may lose environmental services such as bio-diversity, flood regulation and drought control. In any given year, droughts, plagues, floods, hurricanes, extreme storms or violent conflicts affect 5-10% of hungry people.

Meanwhile, global warming can adversely affect weather patterns for farmers dependent on rain. Mechanisms are required to track major ecosystems and their continued ability to produce needed goods and services.

Goal 8: Develop a Global Partnership for Development

The targets are, (a) in cooperation with developing countries, develop and implement strategies for decent and productive work for youth, and (b) in cooperation with the private sector, make available the benefits of new technologies, especially information and communication technologies.

India stands in an advantageous position in necessity-based rural entrepreneurship, but lags behind other countries in terms of opportunity-based entrepreneurship (Mathew *et. al.*, 2001). In rural areas, women tend to get into entrepreneurship only after completing their familial requirements, which means that they try to start enterprises only after passing through their prime age. Majority women were found to prefer traditional areas like home scale processing, tailoring and animal husbandry for starting an enterprise. Application of information and communication tools and techniques in starting service-oriented entrepreneurship can contribute to provide a decent and productive work for rural youth as well as help the rural people to get the benefit of latest technological advancements happening worldwide.

Future Roles of Private Extension Services in India to Achieve Millennium Development Goals

Not long ago, serious doubts were raised on the relevance of private extension services to Indian context, particularly on the inability of the poor and marginal farmers to pay for such services, inadequate infrastructure facilities for private extension services to establish their empire, and the possible misguidance by the unaccountable service providers. Now, these doubts have been shifted towards their roles in the changing scenario. Following roles are immanent, if India has to achieve the Millennium Development Goals, thereby bringing improvements in the livelihood of its vast majority of rural population.

1. *Emphasis on Resource Poor and Marginalizing Indian Agriculture*

A common cry about the weaknesses in green revolution is that the technologies have bypassed the resource poor category. There is also a well-orchestrated feeling that the public extension system has concentrated mainly on the large and progressive farmers, thus neglecting the resource poor category. Any attempt to include private extension must try to address this lacuna of the public extension system by concentrating on the underprivileged group.

Majority of the below poverty level population of India are from rural areas. Even in rural areas, marginal farmers and landless category dominate the below poverty level population.

During 1980 to 1990, proportion of marginal holdings increased from 56.4 to 59.4 per cent, but at the same time, the proportion of larger holdings decreased from 2.4 per cent to 1.6 per cent. While the area under marginal holdings increased from 12 per cent to 15.1 per cent, there was alarming decline in area of larger holdings from 23 per cent to 17.3 per cent (ICAR, 2001). It is obligatory for the private extension service providers to target their prime resources on marginal and small farmers for creating a mass effect.

Achievement of Quality Up-gradation Programme implemented by Krishi Vigyan Kendra (KVK, managed by UPASI–United Planters Association of South India) in Nilgiris district is a model experiment for what could be done to small and marginal farmers. The programme has helped 43100 small tea growers and 21 brought tea manufacturers to achieve spectacular changes in the quality of tea, thereby enabling the factories to pay higher price for small tea growers (Ramu et.al., 2004).

Providing service support for livestock management in remote areas, not reached by public extension system in Erode and Trivandrum districts of Tamil Nadu and Kerala states, has been achieved through Para-technicians trained by KVK MYRADA and KVK Mitraniketan (NGO KVKs) in these districts, respectively (Gowda and Samanta 2002).

2. *Meeting the Emerging Support Needs of Farmers*

A general concern is that the existing extension efforts have placed greater importance to transfer of production enhancing technologies and lesser emphasis to management of farm from the sustainable point of view. The Indian extension system has evolved from its education/teaching mode of 1950s & 60s, and input supply mode of 1970s and 80s to participatory mode in 1990s. The emerging private extension must stand upto the need of the situation by giving something different from what has been offered by the public system.

Some of the emerging needs of the farmers are (Rasheed, 2003);

- What technological options can be used profitably in a given situation?
- How to manage various technologies?
- How and when to change farming systems?
- For which type of products is there a good demand in the market?
- What are quality specifications for the produce and how to achieve them?
- How, when, where, and under which conditions to buy inputs and sell products?
- How to make decisions collectively on resource use and marketing?
- How to find quickly the most relevant and reliable knowledge and information?
- What are the feasible off-farm income generation options?
- What are going to be the implications, if input subsidies are phased out and or if the trade in agricultural products is liberalized?

To ensure sustainability of pluralistic extension systems, building capacity is essentially felt that extension providers should have access to a small but highly qualified and efficient support unit that can carry out prospective studies related to R&D and markets (product characteristics, prices, opportunities).

The provisions that allow the participation of private input and information service providers in the premises of *Raitha Sampark Kendras* (Farmers Contact Centers - managed by the public extension system of Karnataka state) is a welcome change that try to help farmers in getting some of their above needs solved under one roof. It is high time, such efforts are intensified in the interest of small and marginal farmers.

3. *Emphasis on Women and Groups*

Groups add to the social capital of a community. The Social Capital has been defined by Woolcock and Narayan (2000) as the norms and networks that enable people to act collectively. Present individual-centered and progressive-farmer oriented extension approaches ignored this vital capital and in turn remained aloof from the community as such. Besides, the resource-poor farmers and farming situations were not given adequate attention at the required time and frequency. Apart from the high cost, the uniformity of attention has not been paid to deliver messages as per the needs and interests of the hugely varying farming population in its message receptivity.

Representatives of grassroots farmers' organisations cited three institutional and rural contextual variables as most crucial to their emergence (ISNAR, 2000):

1. Public sector research and extension had not addressed local farmers technological demands
2. Farmers realized that the best way to get their needs met was to act collectively
3. A service NGO or other organisation was present to help them organise.

So, farmers have found groups as a better alternative to receive the technological and material support required to practice their profession. With the globalisation and liberalisation of the economy, opportunities for the unskilled and illiterate are not increasing fast enough as compared to rest of the economy. This is leading to an unequal growth in the economy. It is in this context, the organisation of resource-poor farmers and women in rural areas into Self Help Groups has a significant role to play both from socio-economic and sustainable point of view (Gowda et.al.,2003).

Indian farmwomen were not targeted by the transfer of technology and development agencies as they were less accessible and also because their potentials were underrated. Accessibility of the farmwomen for transfer of technology efforts was limited due to the socio-cultural set up prevalent among most rural communities in India. With the exception of Kerala and Tamilnadu states of India, less than 10 per cent extension staff in their development departments is women. This is another reason why the farmwomen were not reached as frequently as the men farmers were reached. Networks, times when they can best be reached, communication media and certain topics are often specific to women or for certain categories of women, in industrial countries as much as in others (Roling, 1988).

Hence it is imperative that the emerging extension approaches take cognacence of this vital fact and devote much of their efforts towards group and women based extension.

The experiences of KVK Gadag, managed by the KH Patil Agriculture Science Foundation in Karnataka state set trend for others to follow in the area of women and group based extension approach. The KVK is working through 1233 groups located in 93 villages, covering 13933 members, for transferring agriculture and rural development activities. The Self Help Group Markets (SHG-Markets) meant exclusively for SHG members, mainly the women, have led an extraordinary success in promoting rural entrepreneurship (Adapur et.al., 2004).

4. *Emphasis on Community Approach to Tackle Environmental Issues*

Participatory approaches are gaining momentum, especially in the area of natural resource management and entrepreneurship development. As the land-man ratio is small, soil and environmental degradation has occurred faster than anticipated. In an effort to produce more from the smaller holdings, the traditional practices contributing to sustainability have been ignored. Therefore, sustainable agriculture, with emphasis on resource conservation, biomass production, crop rotations, eco-friendly practices, and combination of enterprises are high priority extension goals. Resource management, restoration of environmental support and building local facilities will be handled quickly, efficiently and economically, if the people concerned are sensitized, enthused and guided. Hence organising people for development will be a major future extension task (Dwarakinath, 2001).

Conservation of soil and water through watersheds and community participation is being done with financial support from government. Separate line departments have also been created in some states to intensify the efforts with due emphasis for coordination among independently functioning line departments and local governance.

It is quite obvious that the alternate extension approaches address this challenge in a way that adds to the effectiveness of these successful efforts and replicate in larger areas. The MYRADA KVK in Erode district has been doing pioneering job in watershed management through Self-help Affinity Groups (187) and Watershed Development Associations (38) in implementing watershed development activities in about 11932 ha area spread over 21 watersheds (Alagesan et.al., 2004).

5. *Utilisation of all Available Communication Channels*

Use of a particular type of communication channels

may satisfy and effectively meet the requirement of few sections of the rural population only.

World over, the internet spread faster than almost any previous technology. It took radio almost forty years to reach an audience of 50 million and for the internet just four. But, in India, only 3.8 per cent people have telephone mainlines, 0.6 per cent own mobile and 0.68 per cent are internet users. Within this category there is wide gap between rural and urban users, with urban users hugely outnumbering the rural users. However the e-chaupal experiment of ITC in 21000 villages (India Today, 2004) confirms that, modern information and communication tools may not remain as an urban technology.

Print media is probably the only link to outside world for many societies located in interior areas. However in these places, the extent of literacy may still be far below than the national average of 58 per cent literacy for the people above 15 years of age. Among the literates, women lag behind their male counterparts (only 46.4 per cent literacy at the national level). The private extension service providers must take this as a challenge to meet the information requirement of literates located in remote area and to encourage them to co-educate their illiterate counterparts in adopting latest technologies.

Among the mass audio-video media, the introduction of television has to some extent sidelined the importance of radio. However, both these channels are mainly being used for entertainment rather than educational purpose aimed at agriculture and rural development. Both these media have got a vital role to play in educating the rural population to bring about changes in awareness, attitudes, knowledge and skills.

The lesson is loud and clear—use all the means available, both the new and the old, to change information flows and thus enable rural communities to access, share and exchange experiences in sustainable agriculture.

6. *.Coordination among the Private Extension Providers*

There is a strong need to consolidate the efforts to coordinate different types of service providers. Given the large number of independent and often small organisations involved, transaction costs of coordination can be high, but it is essential. Efforts to build multi-agency partnerships in ways that (for the state) reduce costs, but also spread the reach of extension to areas where a purely public sector service is unlikely to be viable, and make it more responsive to local needs and opportunities.

7. *Coverage and Capturing Benefits through Collaboration*

Each type of organisation has its own niche and weaknesses.

- The private sector is focused on the commercial end of the market, unless services are publicly funded.
- Rural Producer Organisation extension services also function best in a commercial environment and may not be effective for poor farmers unless they receive public or donor funding to extend services to poor farmers to unfavorable production environments.
- Non Government Organisations may be best equipped to serve poorer and more marginal groups but may have problems sustaining services that require external funding.

Thus, it is mandatory for all the private extension service providers to operate collectively for the common good of the common man. No doubt, it is more challenging than all the challenges put together.

8. *Ensuring Quality Extension Services*

There are currently few quality controls for service providers. Even where private providers are contracted to deliver public extension, there is often little follow-up on monitoring and evaluation. Some system of quality assurance for the advisory services would be desirable. A mechanism has to be developed by the private extension service providers themselves, so that quality services are offered in the long run, which is essential for their sustainable existence.

Conclusion

Private extension is seen as a mechanism to reduce economic burden of governments, increase efficiency of extension services and increase accountability of extension system. In the changing scenario, it has been accepted as an essential and integral part, to play a supportive role along with the line departments to help the cause of rural population. In the context of Millennium Development Goals aimed at achieving improvements in the livelihood of its population, private extension service providers need to prove their utility in those areas where the public extension system has failed to achieve the desired results. Otherwise, it may end up as another duplicative effort to promote the cause of already reached, instead of reaching the un-reached.

As the purpose of development is to improve people's lives by expanding their choices, the private extension service providers need to adopt different mechanisms, and use all available means of communication. Coordination of all such efforts, big and small, is essential to make a concerted effort to achieve the millennium development goals in desired manner.

References

Adapur, S.H., R. Sudha and L.G. Hiregoudar, 2004. Group Approach for Technology Transfer—Experiences of KVK Gadag. In Eds: S.Prabhu Kumar and M.J.C.Gowda, Innovative Extension Approaches: KVK Experiences. Zonal Coordinating Unit, Bangalore.

Alagesan, P., S. Suresh, K. Arulalan and M. Thangamuthu, 2004. Watershed Development through Community Based Organisations—Experiences of KVK Erode. In Eds: S. Prabhu Kumar and M.J.C. Gowda, Innovative Extension Approaches: KVK Experiences. Zonal Coordinating Unit, Bangalore.

Bhatnagar, S. and R. Schware, 2000. Information and Communication Technology in Development: Cases from India. Sage publications, New Delhi.

Chandrashekara P., 2001. Private Extension: Indian Way. In Ed: Chandrashekara P. 2001. Private Extension: Indian Experiences. Manage, Hyderabad, India.

Dwarakinath, R., 2001. Extension Education in the Context of Changing Agriculture. Leisa India, 3(3): 13-15.

Gowda, M.J.C. and R.K. Samanta, 2002. Paratechnicians as Alternative Veterinary Service for Livestock Management. Leisa India, 4(1): 27-28.

Gowda, M.J.C., S.H. Adapur, and R.Sudha, 2003. Extension through Women and Groups: An innovative approach of KVK Gadag, Karnataka. Manage Extension Research Review, 4(2): 93-115.

ICAR, 1999. NATP Main Document. Indian Council of Agricultural Research, New Delhi, pp. 45-47.

ICAR, 2001. Agriculture Research Data Base, 2001. Indian Council of Agricultural Research, New Delhi.

ISNAR, 2000. ISNAR Briefing Paper No. 15, International Service for National Agricultural Research, The Hague, The Netherlands.

Mathew J.M., P. Srinivas, V.G. Malathi and S.Joseph, 2001. Global Entrepreneurship Monitor—India Report 2001. Indian Institute of Management, Bangalore, India.

Ramu S., J. Durairaj, G.Ramamoorthy, R.Shanmugam and B.R.K. Shivakumar, 2004. Quality Upgradation among Small Tea Growers and Processing Units—Experiences of KVK Nilgiris. In Eds: S. Prabhu Kumar and M.J.C. Gowda, Innovative Extension Approaches: KVK Experiences. Zonal Coordinating Unit, Bangalore.

Rasheed S.V., 2003. Innovations in Agricultural Extension in India. Sustainable Development Department, Food and Agriculture Organisation of the United Nations (FAO). http://www.fao.org/sd/2003

Roling, N., 1988. Extension Science: Information Systems in Agricultural Development. Cambridge University Press, Cambridge, Great Britain.

Shaji, T.K., 2001. India crucially awaits agripreneurship. Agriculture Today, 5(2): 10-12

UNDP 2003. Human Development Report 2003—Millennium Development Goals: A compact among nations to end human poverty. UNDP, Oxford University Press, New York.

Woolcock, M. and D. Narayan, 2000. Social Capital: Implications for development theory, research and policy. The World Bank Research Observer, 15 (2): 225-249.

World Bank 2002. Extension and Rural Development—Converging Views on Institutional Approaches. Workshop summary, November 12-15, 2002, Washington D.C.

Cyber Extension: An Alternative Approach in Agricultural Extension-Some Experiences

Sharma, V.P and Lakshmi Murthy

ABSTRACT

ICTs have started to make their presence felt in Rural India. The farmers and farm-families are browsing the net and getting general, technical and marketing information from the Information kiosks in over 5000 villages across the country. If the rural India can be connected and the "masses" are empowered with "Information", the Indian economy will take a leap forward into the digital millennium with a great speed and the process has already started. The focus on "e-Governance" and "IT for Masses" is also emerging as front-runner in all these State vision documents. Technologies specially suitable for rural areas are being developed and deployed. Portals on rural markets and agricultural services are being hosted. District level Web Sites are being hosted, Information Kiosks are being established at block/mandal and village levels and the technical and other need-based information is being collected, digitized and hosted on the Internet. National Institute of Agricultural Extension Management (MANAGE), Hyderabad has taken up a number of innovative projects to provide information and communication connectivity to the farmers and farm families in rural areas, under the banner "Cyber

Extension". MANAGE has connected over 20 districts, 400 blocks on internet, has implemented Wireless in Local Loop Technology in agriculture, connected over 40 national level institutions on Video Conferencing and providing Video Conferencing access to Farmers' groups and farm-families in rural areas through its Mobile V-SAT Van. The Cyber Extension is not the replacement of existing face-to-face extension mechanism, it is only a supplement to make the existing mechanisms more effective and economic. The ICTs, more dynamic instrument of continuous 2-way dialogue with the farmers on various issues, including agricultural marketing and other forward linkages can help the agriculture extension. All these experiences have been narrated in this paper along with the lessons learnt.

Introduction

A silent revolution is taking place in the communication systems in Rural India. The farmers and farm-families are browsing the net and getting general, technical and marketing information from the Information kiosks set up by a number of pioneers across the country. The total coverage under such initiatives may be very small (around ten thousand villages out of over six lakh Villages in the country), but the potential of this "BRIDGING" of so called "DIGITAL DIVIDE" is being hotly debated within and outside the country. If the rural India can be connected and the "masses" are empowered with "Information", the Indian Economy will take a leap forward into the Digital Millenium with a great speed. The process has already started. The Government of India has already put a "IT Policy" in place in 2000, almost all the states have developed "IT Strategy" and have put these up on their web-sites. The focus on "e-Governance" and "IT for Masses" is also emerging as front runner in all these State vision documents.The industry is also looking at Rural Applications as the potential "Business Domain". Technologies specially suitable for Rural Areas are being developed and deployed. Portals on Rural Markets and

Agricultural Services are being hosted. District level Web Sites are being hosted, Information Kiosks are being established at block/Mandal and village levels and the technical and other need-based information is being collected, digitized and hosted on the Internet.

Agriculture continues to be the occupation and way of life for more than half of Indian population even today. Sustainable prosperity of this class - the farmers, the landless agricultural laborers holds the key for improving the overall Human Resource Development scenario in the country. Indian Agriculture had been on traditional lines till the first waves of Green Revolution in late 60s. The Green Revolution gave a sudden boost to the production and productivity of major cereals in the assured irrigated areas- The Punjab, Haryana and Western U.P. in north, and Godavari, Cauvery deltas in the south. Quick dissemination of Technological information from the Agricultural Research System to the Farmers in the field and reporting of farmers' feedback to the research system is one of the critical inputs in Transfer of Agricultural Technology. The information and communication support during last 55 years has mainly been conventional. The extension personnel of the Department of Agriculture disseminated the technological messages to the farmers manually. This approach has not been able to reach majority of the farmers who are spread across the whole country. This gap remains a challenge for the Extension system even today. To reach over 110 million farmers, spread over 500 districts and over 6000 blocks is an up hill task. The diversity of agro-ecological situations adds to this challenge further. The success of Green Revolution was mainly achieved due to concerted homogeneous extension approach for the assured irrigated area. Now as we move to address the needs of rainfed eco-systems, the extension strategy becomes more complex. Farmers' needs are much more diversified and the knowledge required to address them is beyond the capacity of the grass root level extension functionaries.

Today it is possible to find a solution to this situation by using the potential of Information and Communication technologies to meet the location specific information needs of the farmers. The information and communication networks are expanding very fast. The number of Internet connections in India has crossed the two million mark and the number of telephone connections is over 35 million. The Internet connectivity has touched almost all the districts in the country and is moving down to the block and Mandal levels. Pilot projects to connect rural community to the cyber-space are underway at various locations. The initial response of the rural people, particularly women, has been very encouraging. MANAGE has established Internet connectivity in 28 Districts in 7 States, namely Andhra Pradesh, Bihar, Himachal Pradesh, Jharkhand, Maharashtra, Orissa and Punjab under National Agricultural Technology Project (NATP). Over 200 blocks have already been connected on Internet under NATP.

A number of State Agricultural Universities, government departments and also some private entrepreneurs have hosted Agricultural Web Sites. The Agricultural Information Base on the Net, is building up slowly but surely. MANAGE has taken the initiative to provide linkages to the technical and other farmer friendly information through its Web-Site. MANAGE is also supporting a number of Agricultural Universities and other research and training organisations, both in public, and voluntary sector in building their capacity to digitize the Agricultural Information and host the same on the Web. The web-sites of 4 Regional Extension Education Institutes (EEIs), 6 State level Management and Extension Training Institutes (SAMETIs), 24 Districts, and many others organisations have been designed, developed and hosted by MANAGE. The web-sites of 24 Districts (Agricultural Technology Management Agency, ATMAs), contain very important information on District Profile, Land Use Pattern, District Agriculture scenario, Strategic Research and Extension Plans (SREPs), replicable success stories, and information on important contact persons with their telephone numbers and e-mail-ids. These web sites have improved the information dissemination of these institutions significantly.

Improving Communication Capacity of Agricultural Extension System

The weak linkages among extension, research, marketing network and farmers limit the effectiveness of research and extension to contribute to agricultural development. The Government of India has identified this problem, and is addressing it through national programmes. The erstwhile emphasis on usage of vernacular press, radio and television for reaching to the farmers is being augmented with use of state-of-the-art communication technologies such as Internet and satellite communication. Under the new initiative of NATP, adequate attention is being paid to provide ICT connectivity down to the Block level. This connectivity will facilitate two-way communication among all the stakeholders in the Research-Extension-Marketing-Farmers loop. Apart from core Information and Communication Technology (ICT) connectivity, other forms of audio and visual communication like Satellite Communication (SATCOM) are also being promoted on project basis.

Research-Extension Communication in India: the Present Scenario

The major medium of communication among the research and extension agencies in India is still "Face-to-Face Communication". The DAC (Department of Agriculture and Co-operation)-ICAR(Indian Council of Agricultural Research) interface at the highest level provides the interaction platform for the policy makers in Department of Agriculture and Cooperation and the Senior Scientists of ICAR. Thus, the information and communication support during last 50 years has mainly been conventional. The extension personnel of the Department of Agriculture disseminated the technological messages to farmers manually. This approach has not been able to reach majority of the farmers who are spread across the whole country. This gap remains a challenge for the extension system even today.

State Level: At the state level similar mechanism exists in allmost all the states, where the State Departments of Agriculture interact with the concerned State Agriculture Universities, Research and Extension Scientists to prepare plans for extension activities at field level. In some of the states, the SAUs have established extension centres at each district (Andhra Pradesh - DAATTC (District Agriculture Advisory and Transfer of Technology Centre of ANGRAU). There are few other mechanisms in between these systems. These include the interactions among research scientists and extension managers at Zonal Research Station Level and at Krishi Vigyan Kendra Level. Thus the research extension communication is mainly interpersonal, which is generally followed by technical literature from university side and field feedback (in written form) from extension side. This mechanism prevails in most of the states in India. Under NATP and UPDASP (Uttar Pradesh Diversified Agricultural Support Project) Internet based connectivity has been provided to the major stakeholders in research and extension systems in 46 pilot districts.

The basic communication flow consists of periodical reports to the project agencies and in some cases email communication between researchers and extension managers. The Project districts that have hosted their websites provide basic information about district, the agricultural pattern of the district, the strategic research and extension plan (SREP) and major initiatives under concerned projects. The universities have also hosted their websites and in some cases the ZRSs and KVKs have also done so. These websites are primarily serving as a ready reference for contacting the concerned agencies. They provide contact address, email address and telephone number of important scientists. The technical material on the university websites is of very general nature. The material is however very useful for the university scientists placed at ZRS and KVK levels and extension personnel at the district level. This can be considered as the phase-1 of using information communication technology for agricultural extension. The second phase of this initiative will include e-enabling the research extension communication process in dynamic sense. This will involve re-engineering of existing communication mechanism at various levels.

Defining Cyber Extension Cyber

According to Oxford Dictionary the word Cyber means, "relating to Information Technology, the Internet, and virtual reality, the *Cyber Space*"

The Cyber Space

Cyber Space is the imaginary or Virtual space of computers connected with each other on Networks, across the globe. These computers can access information in the form of Text, Graphics, audio, video and animation files. Software tools on networks provide facilities to interactively access the information from connected servers. The cyber space thus can be defined as the imaginary space behind the interconnected telecommunications and computer networks, the virtual world.

Extension

Extension stands for "the action or process of enlarging or extending something". It could be extension of area, time or space.

Agricultural Extension

Agricultural Extension, according to Dr. D. Benor, " relates to the process of carrying the technology of scientific agriculture to the farmer in order to enable him to utilize the knowledge and a better economy. Agriculture extension service seeks to impart the necessary skills to the farmers for undertaking improved agricultural operations, to make available to them timely information on improved practices in an easily understandable form suited to their level of literacy and awareness, and to create in them a favourable attitude for innovation and change" (Benor et. al 1984). The Extension is an on-going process of getting useful information to people (the communication dimension) and then in assisting those people to acquire the necessary knowledge, skills and attitudes to utilize effectively this information or technology (the educational dimension). Thus Extension is the central mechanism in the agricultural development process, both in terms of technology transfer and human resources development (Samanta, 1993).

Cyber Extension

Cyber Extension thus can be defined as the "*Extension over Cyber Space*". As the word Extension is subject-neutral, so is Cyber Extension. But in the applied context of Agriculture, Cyber Extension means "*using the power of online networks, computer communications and digital interactive multimedia to facilitate dissemination of agricultural technology*". Cyber Extension includes effective use of Information and Communication technology, national and international information Networks, Internet, Expert Systems, Multimedia Learning Systems and Computer based training systems to improve information access to the Farmers, Extension Workers, Research Scientists and Extension Managers.

Cyber Extension: ICT Enabled Approach for Research Extension Communication

The Cyber Extension is not for replacing the existing systems of Communication. ICTs will augment, in big way, the reach and two-way interaction among the key stakeholders. The new technology offers new opportunities. It will add more interactivity. It will add speed. It will add two-way communication. It will add to wider age and also more in-depth messaging. It will widen the scope of extension; it will also improve quality. It will subtract costs and reduce time. It will reduce dependency on so many actors in the chain of extension system, and frankly it will change the whole method of extension in the coming decade. "The continuing rapid development of telecommunications and computer-based information technology (IT) is probably the biggest factor for change in extension, one which will facilitate and reinforce other changes. There are many possibilities for the potential applications of the technology in agricultural extension (FAO, 1993; Zipp, 1994). IT will bring new information services to rural areas which farmers, as users, will have much greater control than over current information channels. Even if every farmer does not have a computer terminal, these could become readily available at local information resource centres, with computers carrying expert systems to help farmers to make decisions. However, it will not make extension worker redundant. Rather, they will be able to concentrate on tasks and services where human interaction is essential—in helping farmers individually and in small groups to diagnose problems, to interpret data, and to apply their meaning (Leeuwis, 1993). The researchers at university now should involve the Krishi Vigyan Kendras, extension functionaries and even the farmers right from the beginning of the project. They can share their objectives, their methodology of research, methodology of analysis and also the observations and intermediate results with the concerned fellow scientists and the interested farmers. In this open system, the other stakeholders can give their feedback and their suggestions to the researchers at every stage of the experiment. In some cases, even the farmers can participate in adaptive research. The validation of research can definitely be done at an appropriate number of locations within the concerned agro-eco-zone, with the results of the same being shared among all stakeholders online, at various stages of research.

The packaging of research recommendation has to be done in more participative way with the help of information and communication technology. The extension functionaries at district level could be taken in to confidence before final packaging of the "Practices" or "Technologies" for each crop. The experiences and results of various trails could also be indicated in the proposed package of practices. The extension functionaries may then keep the concerned researchers informed on the field feedback electronically. This way the ICTs will help both- the researchers and farming community to talk to each other on regular basis.

Reaching the Last Mile: Connecting the Farmers

The concept of "Village Information Shops" is being discussed, debated and experimented in India at various places. Experiments of Dr. M.S. Swaminathan Research Foundation (MSSRF), Chennai, 'Information-Villages' of MANAGE in Ranga Reddy District in A.P., Gyandoot.net initiative of District Administration Dhar, M.P., EID-Parry's Wireless in Local Loop based Village Kiosks in Cuddalore District of Tamilnadu and 'Warna Wired Villages' of National Informatics Centre (NIC) in Kolhapur-Sangli Districts of Maharashtra are some of the cases which provide good insight of farmers and farm-families' information needs and paying capacity. Preliminary results indicate that, 'Agricultural Extension' alone is not sufficient to sustain 'Information Shop' at village or even at Block level. The information supply domain has to be much larger and dynamic so as to offer value-adding information like market prices, local topical information like bus and railway timetables, weather forecasts etc. The experiences of Gyandoot (Dhar) indicate that the 'Village Information Kiosk' can be self-sustainable enterprise (with a potential to provide job for two young rural people at each kiosk), if 'e-governance' services are integrated with the Information Network. The rural people are willing to pay for the information services, provided the services are a little more exhaustive and improve their livelihoods. The 'Extension' information is very important component of the information needed at the village level. The quality and content have however to change quite drastically to make the extension information farmer friendly.

The packaging of Extension Information for the 'Information Kiosk' has to be more visual, more complete (it must provide full knowledge and information about the topic and various scenarios and options to the farmers) and should also indicate the source of information and further reference for crosschecking and clarifications. This will bring in more direct communication between the farmers and researchers and will also improve the quality and language of research-extension packaging and feedback. The lessons from Pondicherry indicate that farmers seek the information on seeds and fertilizers and also on pests and diseases in groups and then they discuss the information at the 'Information Kiosks'. This implies that the information dissemination in the connected villages is likely to happen through the farmers' organisations, farmer interest groups and other informal groups. Another such experiment that has been running over three years now is the "Warna Wired Villages—connected by National Informatics Centre". The backbone for the group access in this experiment was the co-operative society, which has been a very successful phenomenon in Maharashtra since late 60s. In Tamilnadu Nellikuppam Project of n-Logue and EID Parry the basic framework of farmers groups is provided by a sugar factory catchment area. Another technology that is gaining Farmers' confidence is Satellite enabled Mobile Video Conferencing.

Connecting Farmers MANAGE Experiences

MANAGE have taken a number of innovative projects to provide information and communication connectivity to the farmers and farm families in rural areas. We have tried to incorporate the lessons learnt from the other major initiatives, taken up to empower rural India by ICT interventions. Some of the major projects of MANAGE are

1. *ICT Connectivity at District and Block Level under NATP*

MANAGE is involved in implementing National Agriculture Technology Project (NATP) in 7 states and 28 participating pilot districts. Decentralizing decision-making, bottom up planning, multi agency extension delivery and appropriate use of media and ICTs were the major objectives of the project. Roughly 25% of the project funding (to the tune of Rs.30 crores) was earmarked to provide ICT connectivity and content development at 42 project implementing agencies, which include 28 districts (Agriculture Technology Management Agency), Krishi Vigyan Kendras, concerned State Agricultural Universities, concerned SAMETIs and the State Headquarters. The ICT connectivity (computer, printer, modem, UPS, internet connectivity) has been provided at all the PIAs and also at all the blocks (over 400) in the 28 participating districts, Farmers Information Advisory Center (FIACs) have been established at each block and information services on Agriculture, Animal Husbandry, Horticulture, Sericulture and other development departments along with marketing information is being made available at these FIACs. In some states like Maharashtra, Punjab and Bihar, the ATMAs have ties up the information package and delivery with private service providers. Information Kiosks have been opened by private sector for technical information delivery in districts of Ahmednagar, Aurangabad, Amaravathi, Faridkot and Madhubani. The ICT connectivity provided at block level has thus catalysed the public private participation in the extension information delivery at the grass root level.

2. *Use of Wire less in Local Loop (WiLL) Technology to Provide Rural Connectivity*

MANAGE have implemented the WiLL Technology (CorDECT based WiLL Technology developed by IIT, Chennai) at Achalpur Taluk in Amaravathi District under the NATP. A hub with antenna height of 30 meters has been established at Taluk Headquarters which will provide internet and telephone connectivity to a radius of 10 to 25 Kms. Recently at Achalpur major awareness campaigns have been organized jointly by MANAGE, ATMA, Amaravathi and M/s N-logue, Chennai with full cooperation of District Collector, Amaravathi to sensitise the rural community in and around Achalpur Taluk to use this technology optimally. Around 15 information kiosks will be in operatation up to 31st October 2003 and the number is expected to reach 50 by December 2003. Agricultural Information Services along with other rural development programmes is being packaged by a development team constituted by M/s N-logue at Achalpur. The full content is being packaged in Marathi and all the development personnel have been hired from the local area, so as to use the appropriate language for the communication. This is the 1st Wire Less in Local Loop (WiLL) project exclusively in agriculture sector implemented in India.

3. ***Establishing Video Connectivity at Centres of Excellence in Agriculture and Rural Development:***

MANAGE had established a video conferencing studio in August 2000 and used it very effectively and economically to get high quality technical input on agriculture research, education, training and management from national and international organisations. MANAGE has a technical input from FAO, Rome in a national workshop on Cyber Extension in January 2002. Realising the limitation of institutional connectivity in India, MANAGE provided Video Conference connectivity to 25 centres of excellence involved in agriculture research, management, marketing and training. These centres included NIAM, Jaipur, State Agriculture Universities at Jammu & Kashmir, Punjab, Haryana, Rajasthan, Gujarat, Maharashtra, Assam, Uttar Pradesh, Bihar, Jharkhand, West Bengal, Orissa, Tamil Nadu, Karnataka, Andhra Pradesh and Uttaranchal. This provides MANAGE with a strong technical communication backbone among the most important institutions involved in agriculture, research, education and training. The video connectivity among these institutions has triggered a huge increase in inter university communication among the university administrators and scientists. MANAGE has used this backbone to access the best of the resource persons available in these institutions to enrich its training and consultancy programmes. MANAGE has also provided video conferencing consultancy to National Institute of Rural Development (NIRD). It has been connected to six of their sister institutions namely NIRD-Guwahati, SIRDs at Jaipur, Bhuwaneswar, Mysore and Kottarakara (Kerala) along with NIRD headquarters' at Hyderabad. The video conferencing connectivity among these institutions has been found to be an effective communication tool for scientific interactions and also for policy discussions. The video conferencing facilities at these institutions have been appreciated and used by Hon'ble Chief Ministers of Assam, Jharkhand, Madhya Pradesh and so also by President of India in the recent past. Thus the video conferencing corridor in over 40 institutions of national importance in agricultural and rural development is providing face to face communication to over 50% of technical experts in theses areas.

4. *Mobile VSAT Van Based Video Conferencing*

The VSAT based satellite communication on video was available in India since last 10 years. The ISDN networking is now making it available up to state Headquarters and at the most up to district level (not yet covered fully). The rural areas and the farming families continue to remain outside the on-line communications for all practical purposes. To explore the possibility to reach the last mile MANAGE has acquired Mobile VSAT Van based video conferencing which can be established at any place in the country (wherever a four wheeler can go). A Toyota Qualis van has been fitted with a 1.8 meter diameter satellite antenna and with full complement of video conferencing connectivity including a high quality of TV, VC camera, a laptop computer, UPS and generator. The Mobile VSAT Van goes to the villages and the video-conference connectivity is established with in one hour using GPS technology and

VSAT connectivity. MANAGE has send this van to over 100 rural locations in last 3 months (predominantly in Andhra Pradesh). The farmers and farm families from the remote locations have had the interactions with scientists from Agricultural University, Researchers of ICAR system, Trainers from MANAGE and NIRD and policy makers from Government of Andhra Pradesh. The video conferencing sessions have been attended by ordinary farmers, landless labourers, DWCRA group member women, Project staff of DRDAs, project staff of Andhra Pradesh Rural Livelihood Project (APRLP), District Collectors of concerned district, Commissioners of Agriculture and rural empowerment and has also been attended by the Hon'ble Chief Minister of Andhra Pradesh on more than one occasion. All the interactions have been in local languages i.e. Telugu and hence there has been virtually no barrier in communication. Each of the session was attended by around 200 farmers (on an average) from the rural side and by three to four experts from MANAGE side. The impact of this connectivity has surpassed expectations. MANAGE has enabled to reach and communicate with over 20000 farmers in just 3 months and extremely valuable technical discussions have been taken place in this short time. We have now started the process of working out the economics of this project and would like to share our findings with the academic and policy-making community at the earliest.

Thus in India we have over 20 such initiatives where the information services on agriculture and more importantly on agricultural prices and extension have been successfully tried and tested on project basis. All these projects have, however, been isolated attempts and time is now ripe to consolidate the lessons learnt from these initiatives to launch a Nation-wide or at least a statewide connectivity or Information Access project.

Cyber Extension: Getting Started

The process of Cyber Extension in India needs to have a clear vision at national level, state level and more importantly at State Agricultural University level. Learning from the experiences within the country and in other countries, we need to focus on the following key aspects:

1. ***Develop State-of-the-Art ICT Infrastructure to Connect key Stake holders***

    - Providing Telephone, ICT (Internet) and SATCOM (Satellite Communication linkage). Video Conferencing connectivity to all States, SAMETIs, Districts, Blocks, State Agricultural Universities, Zonal Research Stations and Krishi Vigyan Kendras.
    - Providing required hardware and software to all the above institutions.
    - Creating ICT Training and Consultancy infrastructure at National, State and District level.

2. ***Creating ICT Awareness in all Development Departments***

    - Capacity building of all the extension functionaries and scientists at SAUs, ZRSs, KVKs, Districts and Blocks in ICTs.
    - Creating ICT Training Material at National and State level Training institutions.

. ***Create Information Packaging Mechanism at Key Participating Agencies***

- Capacity building of all the extension scientists at SAUs, ZRSs and KVKs in Information Packaging.
- One time digitisation of all extension literature of the concerned SAU/ZRS/KVKs in English and one local language.
- Integration of Agricultural Marketing Data (including Market Prices, Marketing intelligence and state/District specific Forecasts) with extension information delivery.

- Integration of information on allied subjects like Animal Husbandry/Horticulture/Sericulture etc. with the extension information (making it holistic for the farmers).

4. ***Network with e-Governance Initiatives of Concerned State/ District***

- Providing access/links to "e-governance" sites in the district/zone/state.

- Providing links to all general purpose sites to provide area-specific information to the Farmers.

5. ***Create a Nodal Cell in Each State to Monitor the Progress of Cyber Extension on Regular Basis***

- A Nodal Cell with full connectivity with all District and National Agencies to ensure appropriate policy input at State Level.
- Providing technical ICT support to all these agencies for a critical period of one year.

6. ***Identify a National Co-ordinating Agency for "Cyber Extension" in India***

- To provide critical policy support to Government of India and technical back-stopping to all implementation agencies across the country.

- It should be responsible for developing national perspective; a plan of action, getting it approved from concerned agencies and then be responsible for its implementation.

- Workout a mechanism for 'Face-to-face' or telephonic follow-up of 'knowledge-transactions' on monthly/quarterly basis among key Research-Extension agencies within a District/ Zone/State; and

- Monitoring of all 'information transactions' among these institutions and also with extension functionaries/farmers for at least one year.

This process has to be complemented by extensive 'on-line' networking of "Farmer Service Providing Agencies" up to at least block-level. The experiences of NATP have been very positive in terms of providing connectivity at district level, KVKs and block level. Now we need to make this network operational and also expand the same to other states and districts.

Conclusion

The Cyber Extension is not the replacement of existing Face-to-face Extension mechanism, it is only a supplement to make the existing mechanisms more effective and economic. The ICTs can help the agriculture extension, more dynamic instrument of continuous two-way dialogue with the farmers, on various issues, including agricultural marketing and other forward linkages. It is a tool, which can transform the Indian Agriculture from Traditional Agriculture to "Informed Agriculture". All the decisions- right from cropping pattern selection, seed selection, crop-technology, crop-management, post-harvest management, storage and marketing, can become more efficient in line with the consumer-demand and preferences. This is the only way the Indian Farmers can improve their incomes on sustainable basis.

References

Benor et al. (1984). *Training and Visit Extension*, A World Bank Publication 1984. p. 138.

FAO (1993). The Potentials of Microcomputers in support of Agricultural extension, education and training. Rome: FAO.

Leeuwis, C. (1993). *Of Computers, myths and modeling*: The social construction of diversity, knowledge, information and communication technologies in Dutch horticulture and agricultural extension. Wageningan Studies in Sociology, 36. Wageningen: Agricultural University.

Samanta R.K. (1993). *Extension Strategy for Agricultural Development in 21st Century*. Mittal Publications, Delhi.

Zipp, W. (1994). Improving the Transfer and use of agricultural information: Information Technology. World Bank Discussion Paper 247. Washington DC: The World Bank.

Using ICTs in Development: Information Village Research Project, Pondicherry

Senthilkumaran, S., Arunachalam, S., Rajamohan, K.G., Rajasekarapandy, R., Gobu, J. Pakkialatchumy, P. Manikandan, M., Rajamanikkam, R., Sivakumar, P., K. Rameswaran

ABSTRACT

Interest in Information and Communication Technology (ICT)-enabled development is increasing and a number of projects are coming up, but only a few projects have actually delivered. "Only one out of every one hundred telecentres is really useful for the local community where they have been set up, in terms of supporting development and social change", says Alfonso Gumucio-Dagron, development communicator extraordinaire. "There is one thing that we cannot separate from any ICT project in Third World countries: the development of local databases and local web pages that are relevant to the people and that take into account their daily needs, their culture and their language. If this is not embedded into a project, I doubt it will have any positive results for the community. This is why the Village Knowledge Centres in Pondicherry (M.S. Swaminathan Research Foundation) are such an important and coherent experience." While most telecentres that have failed to deliver are like Cadillacs in rural areas, the Swaminathan 'knowledge centres' are like barefoot doctors and the Green Revolution, both of which have delivered and are appropriate to their contexts. In this paper, we will describe the salient features of the project.

Introduction

Interest in ICT-enabled development is increasing and a number of projects are on, but only a few projects have actually delivered. "Only one out of every one hundred telecentres is really useful for the local community where they have been set up, in terms of supporting development and social change", says Alfonso Gumucio-Dagron, development communicator extraordinaire. "There is one thing that we cannot separate from any ICT project in Third World countries: the development of local databases and local web pages that are relevant to the people and that take into account their daily needs, their culture and their language. If this is not embedded into a project, I doubt it will have any positive results for the community. This is why the Village Knowledge Centres in Pondicherry (M.S. Swaminathan Research Foundation) are such an important and coherent experience." While most telecentres that have failed to deliver are like Cadillacs in rural areas, the Swaminathan 'knowledge centres' are like barefoot doctors and the Green Revolution, both of which have delivered and are appropriate to their contexts. In this paper, we will describe the salient features of the project "Information Village Research Project".

The M.S. Swaminathan Research Foundation (MSSRF) chose the imparting of a pro-nature, pro-poor, pro-women and pro-livelihood orientation to technology development and dissemination to the rural and tribal areas as its main mandate when it started functioning in Chennai, Tamil Nadu, South India in 1989.

Beginning

The project was developed in January 1998 by MSSRF as part of its programme for taking the benefits of emerging and frontier technologies to the rural poor. M. S. Swaminathan, the architect of India's "green revolution," is a geneticist by training and an accomplished agricultural researcher. He was aware of the tremendous potential of biotechnology and genetic engineering. In the past few decades, information technology (IT) has emerged as a great force and has changed the way people live and work in the developed countries. Swaminathan hypothesised that if new ICTs could benefit rich countries, why should they not be harnessed to help poor countries.

He convened an international, interdisciplinary dialogue in 1992 with participants concluding that ICTs could have a major role in promoting sustainable agriculture and rural development in the developing world. However, mere technology alone would not do the trick. Wisdom to use the technologies intelligently and innovatively would be essential. The technologies of the industrial revolution have only exacerbated the divide between the rich and the poor. Swaminathan fully realized that all technologies have this weakness. He was looking for ways to harness the benefits of new ICTs while at the same time, preventing them from further exacerbating existing divides. This is a task meant for only a few people and Swaminathan is surely one of the select few.

MSSRF recognized early that the future of food security in the developing world, especially South Asia, was dependent less on resource-intensive agriculture and more on knowledge intensity. The important step in the use of ICTs in sustainable agricultural and rural development is the value addition made to generic information to render it locale-specific. Rural families, particularly marginal farmers and the assetless, can act on such information to improve productivity. It is on this premise that the informatics division started the "Information Village Research Project" in the Union Territory of Pondicherry in 1998, with the support of the International Development Research Center (IDRC), Canada and the Canadian International Development Agency (CIDA).

The project, which started in Pondicherry, was chosen because it had certain initial advantages, such as an accessible government and reasonable telecommunication infrastructure (an urban teledensity of approximately twenty). The level of poverty is high in the rural areas, where 21 per cent of families have less than US$1 per day as family income. The biovillages project, an earlier programme of the Foundation for Community Asset Building based on biological technologies was fully operational in this region. The ICTs project was expected to complement this programme, and derive benefits from the linkages.

The Project

In an experiment in electronic knowledge delivery to the poor, we have set up knowledge centres in 10 villages near Pondicherry in southern India and connected them by a hybrid wired and wireless network – consisting of PCs, telephones, VHF duplex radio devices, spread spectrum and e-mail connectivity through dial-up telephone lines or VSAT – that facilitates both voice and data transfer, which enable the villagers to get information they need to improve their lives. All of the knowledge centres are open to everyone, irrespective of age, sex, religion, caste, and level of literacy and education. No one is excluded. We are providing information on such aspects as crops, market prices, education, health, employment, government entitlements, weather, and fishing conditions. Close to 50,000 people living in these villages benefit from this programme. The knowledge centres are situated in *panchayat* buildings, temples, a midday meal programme centre, and a private house.

The entire project draws its sustenance from the holistic philosophy of M.S. Swaminathan, which emphasizes integrated pro-poor, pro-women, pro-nature and pro-livelihood orientation to development and community ownership of technological tools against personal or family ownership, and encourages collective action for the spread of information and technology.

About half of the population in most of these villages has a total family income of less than Rs 1,250 (about US $25) per month. The project is designed to provide knowledge on demand to meet local needs. The bottom-up exercise involves local volunteers in gathering information, and feeding it into an intranet-type network, which then provides access through nodes to different villages. The 10 villages are connected in a hub and spokes model, with Villianur, a small town 13 km west of Pondicherry, serving as the hub and value addition center. We also look for useful information from elsewhere. For example, we download information on wave heights in the Pondicherry coast from a US Navy web site. Value addition to the raw information, use of the local language, Tamil, and multimedia (to facilitate illiterate users), in addition to the participation by local people right from the beginning, are some of the noteworthy features of the project. Most of the operators and volunteers providing primary information are women, thus giving them status and influence. All of the centers were created because of demands made by the community.

Although we use a wide variety of technologies in this project, our focus is on people and our purpose is to improve their lives through enhanced livelihood opportunities.

The project was participatory right from Day One. The relation between MSSRF and the village community is one of partnership in progress, and not donor and recipient. The entire village community—men and women, landed gentry and the landless, educated and the illiterate—was consulted and involved. We adopted a policy of 'inclusiveness.' Appreciating this aspect of our program, Prof. Bruce Alberts, President of the National Academy of Sciences, USA, says: "My experience in India has made it clear to me that our nation would be much more successful in such endeavors if we were humble enough to incorporate the potential beneficiaries of a service into its initial planning."

We have learnt that we must use the right technology to achieve specific goals. Horses for courses, as they say. We are handling a mix of technologies and the mix has to be smart enough to meet our objectives. We also need to train the local volunteers in their use. Thus we manage social mobilisation on the one hand and technology management on the other, and bring these two together to reap maximum synergy. We have an open mind on technology. In the beginning we started with somewhat high-end technologies such as interconnected computers and communication technologies, but we found even traditional technologies have an important role to play. One thing that needs to be emphasized is that we are not overly keen to use ICTs. This is not our primary goal. Our goal is to empower people through improved access to information. In achieving this goal, if the use of technology provides an advantage, we will use the technology by all means. Initially, we used interconnected computers because we needed a rapid transfer of value-added information from the hub to the villages. For connectivity, we tried the telephone and modem (wherever telephones were available), Motorola two-way radio, and later spread spectrum technology. At the same time we have not abandoned classical technology. For example, we download information on wave height in the Pondicherry coast every day from a US Navy Web site and transmit the information to knowledge centers in coastal villages, where we use the age-old public address system to broadcast the audio file.

Although it is possible to access Internet from five of our knowledge centers, the local communities do not use the facility very often, as it does not provide answers to many of their questions. Indeed we use Internet only for limited purposes, especially for providing weather and wave height information, examination results, employment news and information from Central government departments. Answers to most questions are provided from information collected by our staff and team of volunteers from mostly local sources. Many read the community newspaper (called "Nammavur Seithi") we publish twice a month avidly. We now print 7,500 copies of this free newspaper. The impact of this newspaper has exceeded our expectations. Those who did not know about the knowledge centres came to the centres in large numbers and wanted to use the services provided, especially advertising their products and services. The newspaper made our knowledge centres better known in the entire Union Territory of Pondicherry. Within two days of release of the first issue in February 2002, we received more than 60 calls. Some people told they had only heard about "Kamadhenu" (the mythical cow that gives whatever one wants), but now through M S Swaminathan Research Foundation they obtain development news relevant to the community. Many villagers find our newspaper refreshingly different from the commercial newspapers and magazines that devote considerable space to news about crime and violence, politics, and international affairs. We also use blackboards and billboards at our centers to display the latest news and announcements. We have learned that simply because something is old we need not discard it. And the latest and most advanced technology may not be relevant to local needs.

Many old people have their own traditional knowledge in various fields. But they may not be sharing the information with other people, thinking that the information may not be important for others or assuming that everybody knows it. During the "Nammavur Seithi" content collection we approached village people looking for traditional knowledge, particularly in the area of health. People submitted home remedies for cold, ear pain, pain in veins, fever, etc., pest control methods and symptoms and remedies of animal diseases. Many people told the project staff that they did not know that so much information was available in the village. They found the information very useful and cost effective, as all the resources are available locally. Based on this experience we held several interaction workshops for "Nature Cure and Herbal Remedies". We also contact regularly traditional healers and collect information about traditional knowledge of herbal medicine. So far we have collected more than 600 herbal remedies related to human and animal health.

The users have adapted well to the available technologies. Many of them are now adept at using computers and have enjoyed the benefits of obtaining the information they need. As a result, their information needs are growing. More and more villagers will be adept at using computers and communication technologies, and will wish to take advantage of new technologies. Eventually, we will withdraw from the villages, leaving the entire pro-gramme in the hands of the local community.

Gender concerns are central to the project and we believe that incorporating this concern is essential for project success. Due to a deliberate decision, more than half of the volunteers operating the KCs are women. This has positively reflected on the increase in the number of women users. Handling of PC and answering men's questions give women new confidence and status in the community. In the evening some knowledge centers (KCs) provide counseling to women. Many women have formed Self Help Groups (SHGs) paying monthly subscriptions. They borrow money from the SHGs, when there is a need, especially for education of children and starting cottage industries. The interest on the money borrowed accrues to the SHGs. The KCs help women get training related to new economic opportunities like incense stick manufacturing, pickle making, phenyl and soap oil production and ornamental artifacts from seashells. In the fishing village, there are fewer women users who get news through the public address system. Many women report that they do not have enough time to visit the centre due to the demands of housework and labour. Some women obtain information from other women who have visited the KC.

In the beginning we were not sure how long will the village volunteers, many of whom had finished only middle school education, take to learn to use computers and to operate the associated communication devices. We were surprised at the speed with which they learnt. Remember that they are typing text in Tamil, which has more than 240 characters, using the standard QWERTY English keyboard with 26 characters. As Swaminathan often says they took to new technology as fish to water. Now these volunteers are training others in the villages to learn to use computers!

Creation and updating of relevant content to suit local needs is a key factor in the programme. Prior to commencing content-building activity, extensive consultations were held with participating village communities through small group meetings. It emerged that the provision of dynamic information on prices and availability of inputs for cultivation such as seeds, fertilizer, or pesticides was important to every farmer. Knowledge of grain sale prices in various markets in and around Pondicherry is critical to farmers during the harvest season. This information helps farmers market their produce more profitably. The village residents are most interested in dynamic and customized information. This is a resource-intensive activity and has implications for sustainability in view of the potential of involving more locals to create and manage local and customized information content. An encouraging development in this regard is that some village KCs create content related to agriculture, animal husbandry, education, employment, health, government announcements, income-generating enterprises, the environment, and general information. There is a great sense of ownership among the local people, who do not view the KCs as belonging to MSSRF. This is why there is no vandalism or damage taking place against the property. The community selects the village volunteers. The project staff at Villianur (the hub) and Chennai (MSSRF headquarters) have cordial relations with the local people.

The project to be successful has to be integrative. We mean merely providing information using tools of ICTs to the people is not enough? It has to be accompanied by other activities that will ensure the creation of livelihood opportunities and greater income levels for the community. That is why we facilitate the formation of many self-help groups in each village, liaise with banks for providing microcredit, liaise with training institutions to provide training in setting up and running micro-enterprises, etc. After all information is only one element in the development process. To get optimal results and reap the benefit of synergy we need to integrate it with other key factors.

The project calls for considerable managerial skills. On the one hand we have to mobilize the entire people of the many villages we work in and get accepted by them. That is quite a task. We need to liaise with government officials, local institutions, traders, and a number of others. We need to form self-help groups, help in setting up micro-enterprises, arrange micro-credit, and help market the products coming out of micro-enterprises. Above all, we must work out a smooth withdrawal strategy, ensuring that the knowledge centers, self-help groups, micro-enterprises, etc. will continue to function and with greater efficiency, even after we withdraw.

This project has won two international awards, viz., Motorola Dispatch Solution Gold Award 1999 and Stockholm Challenge Award 2001 under the Global Village Category.

Occasional Problems

Widespread access to information can lead to changes in social equations. For example, many villagers are now aware of government entitlements and they go and ask for them. The officials are now receiving more enquiries and more claims than ever before. In another example, landlords who pay part of the wages to their landless labourers in kind can no longer give lower quantities of whatever they are giving in kind, as the prices are known to everybody. The officials who are bombarded with queries and the landlords may not like what is happening ? There have, in fact, been occasions when certain landlords have brought in labour from outside the village when the local labour demanded the right quantity of rice given as wages in kind. Such situations require considerable tact and persuasive skills to sort out. MSSRF has been successful in tackling such issues. In the early phase of the project, three knowledge centers set up at private residences had to be closed, as the services did not reach all sections of the people in the village and were monopolized by the relatives and friends of the residents.

Often people ask about 'enabling conditions or environment' that can help projects achieve their goals. Honestly, we don't think one should worry about them. In our own case, when we went with the idea of starting this project many villagers, especially the youth, were not very keen to work with us. In fact, we were asked if we were part of the government and if we were going to distribute freebies, and when we told we were neither part of the government nor were we going to give anything free, they lost interest. But we persisted in our efforts and convinced the elders and the women in the village. Eventually the rest of them joined us too. What is important is development agencies should work to create the enabling environment.

As our early surveys revealed, most people in the villages we work were poor and there were hardly any telephone or reading room. If the conditions are favorable, then where is the challenge for the development agency?

Success Stories

Here is what Alfonso Gumucio-Dagron, Development Communicator Extraordinaire, wrote: "For many the criteria for evaluating telecentres until now seem restricted to the financial "success". The bottom line being if a telecentre or radio station makes money, then it is sustainable. No consideration about social sustainability or the impact on social change. Why do we measure social communication projects established to contribute to community development with the same criteria we measure commercial cyber-cafes?

That a telecentre or a community radio station is self-sustainable in terms of funding does not tell us anything about its contribution to social change and development. I do not admire a community based communication project just because it is making money.

Sustainability deals with a wider range of issues. Let us look at ownership, for example: community ownership is key to the sustainability of a community communication project. However, this ownership can have multiple facets. Having a legal title to the facility is one of these, but it is not sufficient to guarantee sustainability. Having managerial responsibility, control over content, and a say in the project's future are equally important. Sustainable community ownership requires that the community has legal ownership, but that also is prepared to take responsibility for the project because it has internalised the sense of ownership. We are happy to report that this is indeed happening now in Pondicherry.

Based on Gumucio's article we are collecting success stories from the villages to know how the knowledge center helps bring about social change. So far we have collected nearly 75 stories. This is an ongoing process.

Two women in Embalam village got the widower and remarriage loans through our government entitlements database and started a dairy unit and provision store. More than 40 village women got the tailoring training through our bulletin and also started a few tailoring units. Some of them got government subsidies for training as well as starting units. A few women got the girl child deposit amount from the government. Many women formed self-help groups through village knowledge centers and started several micro enterprises like incense stick manufacturing, soap oil production, phenyl production, ornamental artifacts from shells, pickles production etc. Below poverty line database also helps the SHG members to get the subsidy from District Rural Development Agency. Embalam village volunteers applied under the "Swarna Jayanthi" scheme with a write-up of what they have been doing in the last four years in the Knowledge Center. What is more important is that these volunteers also persuaded Eachankadu self help group to apply for the same scheme. Incidentally, it was the Embalam volunteers who helped form self help groups and micro enterprises in Eachankadu. In February 2003, two-information village knowledge center groups in Embalam and Echankadu SHG were selected under this scheme and each got a grant of Rs. 3.75 lakh. After winning the grant, they met the Chief Minister (CM) and asked for permission for setting up a petrol station. The CM explained to them the difficulties involved and they decided to open a Kerosene shop for five villages. One women volunteer in Kalitheerthalkuppam center formed a thrift society with a group of 200 members.

Through market information several farmers got a good price for their grains from different markets. Tender coconut sellers regularly check the "farmers market" price information. Information provided on various farming techniques and pest control methods help the rural community extensively. Many farmers get the information on inter-cropping and multi cropping and implement it in their fields. Most people are using our animal husbandry database, doctors' addresses and wireless phone (in emergencies) facility. One auto rickshaw driver analysed month long market information and submitted a proposal for supplying vegetables to a beach resort. He got the order and now supplies the vegetables regularly to the resort.

Teachers are using our Educational CDs to improve the science/arithmetic knowledge of children. They also prepare question papers in the computer in the KCs and issue the printed questions to the students. Before that the children had to copy the questions written on the blackboard by the teacher, spending some time on the job. They will see a printed question paper only in the public examination of 8th or 10th Std. The local school headmasters use the center for typing certificates for students (for having studied in the school) and relieving order for teachers. Many children/students/public are getting training in use of computers. Educated youth surf the net and develop market linkages with several companies making milk powder, fruit juice etc. Through Internet many students get the examination results, education course details, software development details, entrance examination results, medical and engineering counseling details, etc. Many college students and research students are preparing their project work in the knowledge center computers. Many students use computer dictionary.

Panchayat (Local body) leaders are keeping the accounts, typing letterheads/letters in the knowledge centers. The resource persons of District Rural Development Agency and Small Scale Industries Corporation of Pondicherry use the centers for preparing course material, training guides, trainer's details, summary of the training etc.

Several people got employment in fire service department, army, private companies, police service, education department etc. after seeing the advertisements at the KCs. Private companies and employee unions are using our system and preparing their accounts, salary certificates, subscription details, employee details, etc.

Many old people share their traditional medicinal information and make it available for all the networked villages. Through the centers the villagers have got many benefits needs like roads, streetlights, drainage facility, bus facility, and compound wall for cremation place.

Weather report is very useful for farmers, brick manufactures, fishermen etc. If somebody wants to contact a person in the village they call the village knowledge center phone. When a government official receives an application from the villagers, if he/she needs any clarification, the official uses the wireless telephone at the KC to contact a higher official to be able to take an immediate decision. This is appreciated both by the official concerned and the applicant. Based on bullion and silver prices provided at the KC, people buy gold (whenever the price is low). Many pregnant women benefited from the medical care information about pregnancy. A few people got STD/ISD code information from the village knowledge centers. In the morning, many villagers are visiting the center to read the daily newspaper. LIC agents are using Internet to find out the various income tax schemes, commission, and also submit their yearly reports to the clients using computer/ printer facility. Govt officials use the knowledge centers extensively; they type their letters and reports, and issue caste certificates, monthly income certificates, voting cards, ration cards etc.

Sharing of Experience

In October 2002, we held an eight-day Workshop on "ICT-enabled Development: South - South Exchange". The workshop was supported by the Humanist Institute for Co-operation with Developing Countries (HIVOS), The Hague, The Netherlands; International Institute for Communication and Development (IICD), The Hague, The Netherlands; and International Development Research Center (IDRC), Canada. Twenty-one participants from ten countries (India, Mongolia, Malaysia, Philippines, Sri Lanka, Tanzania, The Netherlands, Uganda, Honduras, and Zimbabwe) spent three days at MSSRF, Chennai and the remaining days in the Union Territory of Pondicherry and Kannivadi region (Tamil Nadu), where MSSRF have our knowledge centers to serve the local communities. The exchange programme included participants' visit to the rural knowledge centers, discussion with the community on how to strengthen the multiple livelihood opportunities through information and communication technolgoies and interaction with key government officials.

The objective was to learn from one another, be stimulated by a good example and distinguish which experiences could be replicated as they are and which ones needed to be adapted to the local situation. The participants had ample opportunities to interact with our village communities and volunteers and to exchange knowledge and experiences with them. This was one of the most rewarding experiences.

In March 2002 MSSRF joined hands with One World International in an experiment on creating an Open Knowledge Network, to test the possibility of connecting villages in different continents using world space radio and exchanging information—both uploading information from each centre to one of the world space satellites and downloading the information on to a world space radio and jacking it up to a PC where using XML filters one can download only the information relevant to a particular village or community. Subsequently One World has carried out two pilots, one in East Africa and another in West Africa. Soon all three groups will work together in exchanging information. A large number of researchers, students, experts from international agencies and journalists have visited the knowledge centers and written about this work and invited to talk about the work at many national and international conferences and workshops.

Lessons Learnt

The programme has been designed on the *Antyodaya* (unto the last) principle of Mahatma Gandhi, i.e., ensure that the poorest person in the village gains from the technology and that technology does not further enlarge the rich-poor divide. Some of the lessons we have learnt are the following.

- We owe our success to the vision of our leader, Prof. M S Swaminathan. He thought we could use ICTs as a crosscutting theme that can benefit other sectors like agriculture, health and education and the results proved us right. It was indeed a bold decision.
- Connectivity, content and sustainability should receive concurrent attention.
- Constraints must be removed on the basis of a malady-remedy analysis; for example, wired and wireless technologies could be used where telephone connections are not adequate or satisfactory. Similarly, solar power can be harnessed where the regular supply of power is irregular. The approach should be based on the principle that there is an implementable solution for every problem.
- The information provided should be demand driven and should be relevant to the day-to-day life and work of rural women and men. Also, semi-literate women should be accorded priority in training to operate the centre, since this is an effective method of enhancing the self-esteem and social prestige of women living in poverty.
- Knowledge dissemination should be linked to access to the inputs needed to apply the knowledge for economic activities.
- The Knowledge Centres should operate on the principle of social inclusion, there by presenting a win-win situation for all.
- The programmes designed to empower rural families with new knowledge and skills should be designed on the *antyodaya* model, where the empowerment starts with the poorest and most underprivileged women and men.
- The local population should have a sense of ownership of the Knowledge Centre. It should be client managed

and controlled, so that the information provided is demand and user driven.

- The local population should be willing to make contributions towards the expenses of the Knowledge Centre, so that the long-term economic sustainability of the programme is ensured. Contributions in cash or kind generate a sense of ownership and pride and create an economic stake in the operation of the centre.

Virtual Academy for Food Security and Rural Prosperity

From this experience MSSRF has gained some useful insights on how access to information and the use of ICTs can help in poverty eradication and empowering women. The bottom up project, unlike many telecentre projects, is an integrated development project in which we try to use ICTs wherever necessary in every aspect of development. We are now going one step further and have established a Virtual Academy for Food Security and Rural Prosperity (VARP).

The VARP is designed to assist in ushering a knowledge revolution in rural India through the effective and meaningful use of modern information and communication technologies. The academy will enable the farmers' organisations and village women to easily access scientific and technical knowledge that they need to solve local problems and enhance the quality of their lives, as well as to communicate their own insights and needs back to scientists. The main aim of the Virtual Academy is that knowledge should reach every home and hut.

This Academy will help to spread demand driven and user-friendly information through the integrated mobilisation of the Internet, cable TV, community and ham radio and vernacular newspapers. To honour community volunteers who have demonstrated a high degree of commitment we have introduced the system of electing

'Fellows' of VARP. The Fellows of the Academy, chosen by a peer review process, are semi-literate or literate rural women and men who have shown extraordinary capabilities in mastering new technologies and applying them for the common good of all the families in the village. The first batch of Fellows of the Academy (4 women and 2 men) are from very poor families and are products of the day-to-day struggle of farmwomen and men belonging to small farm families in meeting their livelihood needs. This project has shown that if our vision is clear and if we work closely with the community, we can use ICTs to make advantage in development projects.

Acknowledgement

We are grateful to our donor agencies – IDRC, CIDA, the Government of Pondicherry, Ford Foundation and Friends of MSSRF, Japan—for financial and in-kind support. We are grateful to the village communities in Pondicherryfor welcoming us to work with them.

Strategy Focused Knowledge Management Support for Extension Services—The KISSAN-Kerala Approach

Srivathsan, K.R., Hariprasad and Ajith Kumar

ABSTRACT

Karshaka Information Systems, Services and Networking (KISSAN-Kerala) is a project of the Department of Agriculture, Kerala. The project was conceptualized by the Indian Institute of Information Technology and Management—Kerala (IIITM-K). Kerala Agriculture University, and Farm Information Bureau now currently run the project jointly with IIITM-K. The project is in the pilot stage. The aim of the project is to address farm and agriculture related issues of sustainability, productivity and profitability by providing the farmers and other stakeholders of agriculture the 'Right information at the Right Time to the Right Persons in their Right Places and in the Right Context'. These five 'rights" are called as the 5Rs. The project runs a central Agriculture Data Aggregation and Knowledge Management portal (see www.kissankerala.net) over the Internet. Here the agriculture data from various sources related to market, weather, events, etc. are collected. The same, after due validation process is made available to all openly. Queries from farmers and others may be posted over the portal. Based on the nature of queries, they will be sent over the net to concerned experts in the subject wherever they are within the state or even elsewhere.

Farmers get authentic answers within a reasonable time. Difficult questions are forwarded to the Agriculture University for further study and research. The portal also hosts a rich repository of information on crop index and databases, best practices, information on where to get what, help desk, etc. Farmers may even advertise their produce and at what minimum price they may like to sell. We shall be adding information related to world-trade related information on crops, produce and trade related opportunities of relevance to Kerala. To facilitate the reach of the latest information and as a means of general education, the project scientists produce weekly TV programs broadcast over Asianet channel called 'Krishi Deepam'. The portal services of relevance, best practices and information of strategic value are presented to the viewers in this program. Thanks to the dual approach, the KISSAN portal receives many series queries from across Kerala and outside. A notable feature of the portal is that it has backside support for strategy-focused groups to provide advisories for extension officials and workers. A feature is being added by which any concerned farmer asking a query will be attended by extension services field official with full backing of the relevant strategy group. The group members themselves may be anywhere in the state. This is much like the Telemedicine concept in health sector. There is much scope for supporting dozens of such strategy-focused groups and their services. Each such group may be funded independently to carry out their respective services. Such a concept will lead in future to the emergence of effective agricultural consulting services by experts in their concerned specialisation. This appears to be the best route to introduce large rural employment in knowledge intensive products and services of relevance to agriculture.

Introduction

Till recently India has seen little development and use of IT services in Agriculture. The little use that is going on today is limited in some pockets to provide web-based information, email supported consulting services and one successful case-e-Choupal (see www.echoupal.com)-an industry initiative in farm supplies and trade promotion. Even these have had some significant impact on the farmers. The scope of using IT in Agriculture is much larger than the current attempts. Recently the Kerala Department of Agriculture has approved and launched the *Karshaka Information Systems, Services and Networking, or KISSAN-Kerala,* as a project jointly executed and managed by the Indian Institute of Information Technology and Management — Kerala (IIITM-K), Farm Information Bureau (FIB), Kerala Agricultural University (KAU) and the Kerala Governments Directorate of Agriculture. KISSAN-Kerala has now an active knowledge management portal in operation (visit www.kissankerala.net) and launched a well-received weekly television programme called Krishi Deepam over the Asianet main TV channel. Much has been learnt in the last four months since the programme was initiated. The progress of this was reviewed on December 10, 2003 in a workshop held at the KAU.

Under the KISSAN project, the triad of KAU, Department of Agriculture and IIITM-K are moving together to build and establish such empowering organisational learning practices in agriculture that takes the gains of knowledge management to the farmers. This paper provides an overview of the objectives, approach and the core concepts behind the KISSAN project. It also suggests the directions on how this project may be extended to provide comprehensive knowledge intensive products and services in agriculture, healthcare, education and other services of great significance to India's rural development. In addressing the approach to KISSAN and the Virtual University for Agrarian Prosperity (VUAP) programs and services over it, we set ourselves the following vision, approach and objectives.

1. Our Vision is to enable, educate and empower every stakeholder in Kerala Agriculture.

2. Our approach will be holistic and knowledge management centric with the stated aim of 'Right Information to Right Persons at the Right Time in the Right Places and in the Right Context'. Every concerned group or organisation in agriculture will be involved.
3. Our methodology professionally addresses the concerns of sustainability, productivity and profitability of the farmer and agriculture related industry and commerce. The actions in these will be aligned with concerns for ecology and environment and balanced against the emerging globalisation of trade in agriculture related goods and services.
4. Our goal will include generation of high quality agriculture technology advisors and consultants as part of the KISSAN/VUAP programs.

The project infrastructure provides excellent infrastructure and organisational support for establishing IT facilitated Agricultural Consulting Services through a virtual enterprise model. It is a model where cooperating experts who may be located anywhere geographically will be in a position to offer specialized and advanced consulting services or help build and sustain Knowledge Intensive Products and Services in the rural areas. The services may be offered anywhere in the world. Beginning with Kerala as the Pilot, KISSAN will also strive to take the project to the national level.

The Setting of Kissan-Kerala Project

The project has developed several innovative IT facilitated methodologies that assist the Agriculture Extension related services. These include collection, validation and presentation of market and other information of changing nature that are of importance to all the stakeholders in agriculture. The project supports knowledge management practices supported through a central

Agricultural Data Aggregation and Dissemination Portal that is located in the Agriculture Data Centre at IIITM-K. The different components of KISSAN activities are given below.

(i) Development and maintenance of an interactive Knowledge Management Portal www.kissankerala.net as the central part of the KISSAN Agriculture Data Centre (KADC) located at IIITM-K. The portal services are made available to all stakeholders in Kerala agriculture — from scientists in the agriculture research institutions, university scholars, FIB and Government officials, related persons in trade and commerce organisations to the field level extension officials and other government officials.

(ii) Establish rural information Kiosks in the Krishi Bhavans for farmers to access the portal and interact with concerned services groups. Initially the Kiosks are being set up in the Trivandrurn rural district area on a pilot scale. However, the services are accessible through any internet access points like the home/office internet connections or cyber cafes.

(iii) Development and management of role-specific content-both static and dynamic -posted in the KISSAN Portal for use as agriculture advisories. The content development responsibilities are to be distributed and made available to the relevant groups that use the portal for offering their respective services.

(iv) Production and broadcast of agriculture television programs to build awareness, answer queries of generic kind and provide dynamic information like crop advisories, validated and analyzed market information, emerging IP/WTO related issues, etc. This TV programme is now broadcasted as weekly serial called KISSAN Krishi Deepam over the Asianet TV channel.

(v) Making the services of KISSAN available to all interested parties from traders, export promotion bodies, farm inputs supply agencies, scientists, agriculture consultants,extension services, government programs, etc.

(vi) Extend KISSAN services to related sectors like animal husbandry, poultry, fisheries. etc.

The spirit of KISSAN project is to provide holistic integrated information services management support for agriculture in ways that enhance (i) healthy sustainability, (ii) productivity and (iii) profitability to all concerned. The goal is to enhance the competitiveness of Keralas agriculture. The approach is solidly built around a collective IT facilitated strategy focused Knowledge Management framework.

Knowledge Management in Kissan-Kerala

Till recently, Knowledge Management (KM) as a subject has been taught in some elite management schools and practiced in some large consulting and IT enterprises with the support of Enterprise Applications Integration [EIA] or Intranet portals. The reader is referred to (Dataquest magazine, 2003) for the current trends in implementing EAI in the industry. However KM is evolving into a deep and complex subject. Some insight into it may be had from (Srivathsan, 2004). From the KISSAN perspective, we may say that KM is about:

(i) capturing the best practices in agriculture and related management of the different processes.

(ii) disseminating the same to those who will benefit from it.

(iii) re-using the knowledge base gained from the experience of all concerned in ways that avoid

making the mistakes from lessons learnt from elsewhere, and propagate the good practices learnt earlier.

(iv) collaborating in a focused way with those who have similar needs and expertise for enhancing the sustainability, productivity and profitability in the relevant activity.

All the above four activities are enhanced by the kind of IT facilitated processes and management that provide for the 'Right Information to the Right Persons at the Right Time, in their Right Places and in the Right Context'. What these 'Rights' are determined by the subject domain and the concerned experts and collective wisdom of all concerned for the specific problems or issues they address.

The challenge is to provide the kind of IT infrastructure and imaginative uses of mass media like the television that support IT facilitated assistance to farmers by experts. The aim is to provide effective consulting and decision support inputs to the farmers and address the specific issues related to the agriculture related extension problems. The way by which this infrastructure will help the agriculture extension services is best illustrated by the following example.

An Example of a Scenario in IT Facilitated Extension Service

Consider a spice plantation farmer in Malappuram district of Kerala who has problems of productivity related to his pepper crop. The plants appear to be stressed and he wishes to seek the help of agriculture extension service. He may watch the 'Krishi Deepam' Television serial of the KISSAN project and will be aware of the www. kissankerala.net portal and associated services. So he decides to go to one of the nearby 'Akshaya Cyber Cafe and visit the KISSAN portal. Now he may go through the following experience.

(a) Farmer becomes aware that there is a Pepper Productivity Experts Group (let us abbreviate it as PPEG) that provides expert advice for his

specific problem. The members of the authorized Expert Group may be in different places across the state (or for that matter anywhere in the world) so long as they are well connected over the internet to the portal. The different experts may be from the Spices Research Institute, individual agro-consultants, faculty from the Agricultural University, etc. The PPEG is formally constituted and recognized through a registration process.

(b) Farmer becomes aware from the portal that the Agriculture Department has a financing scheme to support measures to enhance pepper production in the region.

(c) He types in his request for help from the PPEG, briefly provides his details and the problem. The farmer if registered in the portal may submit his request for attention through SMS and interact with the PPEG through a cellphone as well.

(d) The portal automatically alerts the PPEG experts through email, or even as SMS to their respective cellphones.

(e) The individual PPEG experts visit the portal and get the message typed in by the farmer. The experts may discuss among themselves over an audio (or even video in the future) conference, chat or study current issues related to the region, get data about soil conditions related to the farm (if available), agro climate of the area, etc. - all through the KISSAN Portal.

(f) PPEG identify the extension official nearest to the farm, sends him/her a detailed instruction to inspect the farm and submit a case report through the portal within a stipulated time.

Farmer is also contacted over telephone if possible.

(g) Concurrently the farmer himself wishes to take some lessons on the topic from the portal (including multimedia lectures) and learn about how the crop productivity may be restored.

(h) The Extension Official visits the farm, may collect digital camera photos of plant conditions. His report, with the findings and photos are submitted through the Akshaya rural Internet access facility.

(i) After some interaction, perhaps a field visit and examination by some of the experts in the PPEG, the problem of the farm may be diagnosed as suffering from some form of disease and a treatment by one of the biocontrollers like Trichoderma verdi or Pseudomonas may be recommended.

(j) Extension Official helps and certifies all papers of the farmer are in order and recommends financial support under the government program. The anti-fungal treatment is done by a professional agriculture services agency in the region.

(k) Farmer also learns through the portal, the TV programs extension based education and the opportunities to enhance his income through future trading and support for export of his crop for profitability.

The above scenario may appear rather simplistic in relation to the ground realities. But it is intended to communicate the picture that strategy focused groups or expert groups with members distributed across geography will be able to provide effective agriculture consulting services through appropriate management of network and media supported IT facilitated processes. We call this model as Virtual Agricultural Consulting Services (VACS). The VACS scenario presented above and sustained through the sophisticated Knowledge Management and Interaction Portal at the Agriculture Data Centre permits practically any number of parallel VACS organisations to be operational. VACS is extendable into a powerful model for supporting agricultural consulting services as a sustainable business model. It provides the true example of how IT can support 'Knowledge Intensive Products and Services', a concept for rural development that is often at the core of the President of India, Shri A.P.J.Abdul Kalam's Vision 2020 presentations (see http;//www.presidentofindia.nic.in/). How the KISSAN project has implemented, the support framework for the VACS using a combination of advanced Knowledge Management Portal, Internet, telecom and TV medium and the implementation approach are briefly explained in the next section.

Kissan-Kerala Information Systems and Services Network

The KISSAN-Kerala project has established an interactive information exchange and collaboration oriented Knowledge Management Portal for Kerala's agriculture. The project has also added TV broadcasting as a method of quick dissemination of best practices, government programs, collection and presentation of market, weather and agriculture related information with effective validation processes. For example it is possible to use KISSAN infrastructure as a base to address trade sensitive agriculture production management that aims at Kerala agriculture becoming a globally respected supplier of quality agriculture products. The KISSAN IT infrastructure that has been established is illustrated in Fig. 1. At the core of the KISSAN network is the KISSAN Data Aggregation Centre at IIITM-K. It hosts an advanced Knowledge Management Portal developed in house by the KISSAN and IIITM-K team. It provides for the Knowledge Interaction Interface for every group that provides powerful support for collaboration tools, digital library, relevant knowledge base and information organisation as suited to the respective groups' functions. The portal is built over an Enterprise KM platform that was earlier developed at IIITM-K under a Technology Incubation program. The core KM server is called Trans-E and it is now developed and serviced by Transversal E Networks (P) Ltd. (see www. transversalnet. Com), the company that was born out of IIITM-Ks technology incubation.

The agricultural scientists and experts who provide the VACS may be located in different parts of the state and access the internal areas relevant to their services through this portal. Internet Kiosks have been established in the Trivandrum district area for the farmers and interested persons to access this portal and post queries, interact with experts or obtain relevant information. The portal services will also be made available to rural Cyber Cafes (like those in Malappuram Akshaya project) for access to citizens anywhere.

Fig. 1 shows health and education as additional services in dotted boxes. It is intended to show that the same network with possibly different data aggregation centers hosted by universities or healthcare organisations may be used by respective agencies to provide Knowledge Intensive Products and Services in health and education using methodologies similar to those in KISSAN. Efforts are already on between the Kerala Agricultural University and IIITM-K with funding from the Department of Agriculture to support the programs under the Virtual University for Agrarian Prosperity (VUAP) for the rural farmers, farm workers and agriculture related trade, consulting, farm inputs suppliers and services persons. Rural areas require integrated services in agriculture, health, education and e-governance. The KISSAN network with requisite management to sustain new concept services like the VACS provides a good functional model for the knowledge intensive products and services. The effective

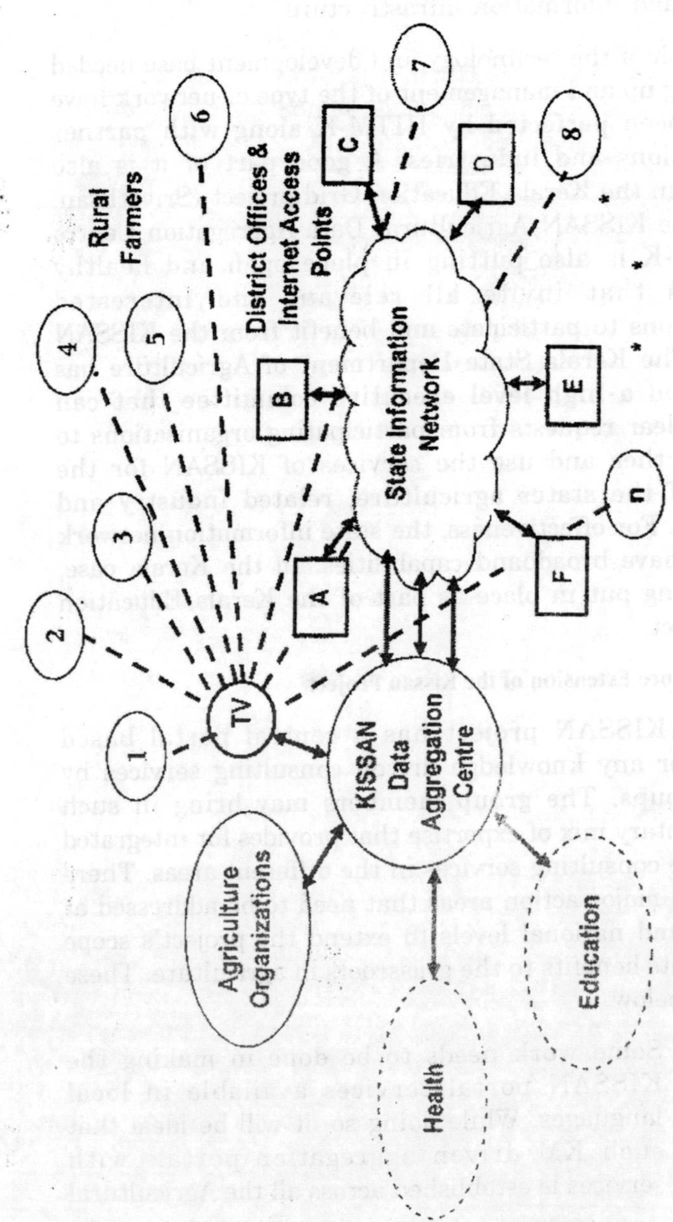

Fig. 1. KISSAN Information and Media Network

cost of running such services will be affordable if all the three—health, agriculture and education—share the same network and information infrastructure.

Much of the technology and development base needed for setting up and management of the type of network have already been perfected by IIITM-K along with partner organisations and industries. A good part of it is also deployed in the Kerala Education Grid project (Srivathsan, 2003). The KISSAN Agricultural Data Aggregation Centre at IIITM-K is also putting in place open and healthy protocols that invite all relevant and interested organisations to participate and benefit from the KISSAN facility. The Kerala State Department of Agriculture has constituted a high level executive committee that can speedily clear requests from participating organisations to come together and use the services of KISSAN for the benefit of the states agriculture, related industry and commerce. For effectiveness, the state information network needs to have broadband capabilities. In the Kerala case, this is being put in place as part of the Kerala Education Grid project.

Issues in Future Extension of the Kissan Project

The KISSAN project has a central portal based support for any knowledge driven consulting services by expert groups. The group members may bring in such complementary mix of expertise that provides for integrated agriculture consulting services in the different areas. There are several major action areas that need to be addressed at the state and national levels to extend the project's scope and reach its benefits to the grassroots in agriculture. These are listed below.

(i) Some work needs to be done in making the KISSAN portal services available in local languages. While doing so, it will be ideal that such KM driven aggregation portals with services is established across all the Agricultural

Universities of India and supported by programs jointly driven by all concerned in each state of India.

(ii) A core need for such services management is the requisite manpower to manage the services. It will require IT savy agricultural scientists. To over come this shortage, the Kerala Agricultural University and IIITM-K have planned to offer jointly a two-year duration M.Tech. Degree programme with different specialisation options in Agri-Informatics for graduates of Agricultural Sciences. Suitable curriculum upgrade to introduce agriculture information systems in the B.Sc. and M.Sc. programs in agriculture will also be taken up.

(iii) For effective administration of KISSAN services and its benefits to reach out to the farmers directly, we need a farmer registration database. This will provide a farmer identity code that links to a (confidential) database having details concerning the GIS map of his landholdings, his plantation or crop related history of practices and produce, agroclimatic information, etc. The farmer may use this identity through SMS over cellphones to ask for assistance even without logging onto the internet. The same identity will help in updating the farmer's database in a coherent way. Such database will also help planning of agriculture related activities.

(iv) As of today, the project is funded by the State Department of Agriculture. There is an urgent need to undertake several IT related developments to build the database of diverse kinds that are needed to enhance the services provided by the KISSAN Project. IIITM-K with support from KAU is preparing a proposal for such a centrally funded R&D Centre.

(v) A major requirement for quality Agricultural Consulting Services is the capacity to build and maintain very large knowledge bases that will have volumes in hundreds of terabytes. Building such very large databases need the involvement of multiple organisations supported by an underneath technology that makes it possible to coherently aggregate, maintain and query such knowledge bases, IIITM-K has already begun studies in this and is developing the concept of a Scientific Portal for Agriculture. IIITM-K and KAU will soon submit project proposal to a relevant central agency like the Indian Council for Agriculture Research, or the FAO.

(vi) lllTM-K is coordinating a major project in Higher Education called the Kerala Education Grid. Recently the project has come up with the concept of Virtual Learning Campus (VLC). The effective benefits of the KISSAN project will be considerably enhanced by integrating it with the VLC concept of the Education Grid. For information concerning the Education Grid and the VLC. The VLC besides enhancing the KISSAN and the VUAP, will bring in high quality education to the rural sector.

(vii) A major challenge in setting up the services at the national level is to provide quality broadband Internet connectivity in the rural areas. The current patch up efforts in setting up rural connectivity in fragments and pockets of the country will not do. The VLC paper (Srivathsan, 2004) provides an approach to build such network. A good strategy is to use the fiber cable network of the Railnet. Indian Railways has OC-3, or 155 Mbps access points virtually in every rural station of significance covering a route length of over 24000 Kms. The Power Grid network is another example, although it does not have frequent taps. A radio mast at each of these taps with 802.11 b or similar wireless access network will possibly cover 10 Kms on either side of the tracks. That is nearly 500,000 Sq. Kms of India's rural area accessible directly! With some repeater hops and hubs, we shall be covering nearly three times as much, i.e., half of India's area reaching most of the continental parts of India. Satellite based networks under GSAT-3 of ISRO may complement this network.

IIITM-K along with KAU is keen to propose and launch an advanced postgraduate education and research institution that is dedicated to the development and management of IT applications and services in agriculture and related areas. Government of India at the Centre must take several steps to facilitate rural development along the above lines. These should be in synergy with the revamping and modernisation of Agricultural Universities, R&D organisations and state level extension services.

Conclusion

This paper presents an overview of the KISSAN-Kerala approach to IT facilitated services in agriculture. This approach is totally fresh and addresses directly the concerns of farmers and other stakeholders in agriculture. It brings in the multiple organisations, the governments, various agencies and agriculture related industry together. It is a holistic approach that brings together the currently isolated and fragmented organisations and agencies in agriculture to work coherently for agrarian prosperity. The methodology adopted has the potential to build the awareness among all stakeholders and empower them to make Indian agriculture globally competitive. We have shown how Virtual Agricultural Consulting Services can be supported and managed over a web-enabled knowledge base through the knowledge Management approach adopted in the KISSAN services. Several major issues related to how IT can be used in agriculture, related industry, trade and commerce are brought out. The case is made that a major postgraduate education and research institution in IT applications and services needs to be set up. IIITM-K itself may be upgraded to become such an institution. Several suggestions on how we can take some major national initiatives along the above lines are provided.

Acknowledgements

The KISSAN-Kerala is a project approved by the High Level Committee for applications of IT in Agriculture appointed by the Kerala Department of Agriculture and chaired by Dr. V.L. Chopra. Shri K.P.P.Nambiar who is also the Chairman of the Board of IIITM-K, chaired the IT part of the committee. The authors acknowledge each and every one associated with the KISSAN and related efforts for making it possible.

References

Dataquest Magazine. Oct. 2003 Issue with special coverageon EAI.

K.R. Srivathsan. 2004. "The 'I' in IT — a Pancha Kosha View", Jan. available for private circulation. Being submitted for publication.

K.R. Sriyathsan, 2003. "Management of Refereed Content Generation and Utilisation in Formal Higher Education". Global Journal of Flexible Systems Management, Vo 4, Nos. 1 & 2, January — June.

K.R. Srivathsan. 2004. "Future ICT Infrastructure for Education". Jan., privately circulated paper.

ITC e Choupal

Sudipta K. Mukhopadhyay

ABSTRACT

ITC *e choupal* is a unique web based initiative operating through kiosks at villages, which is managed by a lead farmer appointed by ITC named "Sanchalak". The reach of this network in the remotest of the villages and its easy two-way communication flow acts as an efficient and effective method for technology transfer in the villages. Each *e choupal* has about five or more villages in its encatchment. The infrastructural bottlenecks of power and telephone connections are overcomed by providing solar panels and VSAT's at the *e choupals*. The Sanchalak is properly trained to deliver the information and advice provided at the choupal to the farmers. It acts through crop specific portals for different regions of India and offers a plethora of other advantages aimed at raising the well-being of the farmer by making him take an informed decision. The web-based initiative is augmented by frequent visits of ITC people who conduct village meetings, address issues and appraise the villagers of the offerings through *e choupal* and take continuous feedback. ITC has collaborated with many organisations and universities to provide contemporary technical expertise, quality inputs, better farm management practices and customized information at the *e choupals* on a continuous basis.

Introduction

ITC's diversified business portfolio has enabled the company to create and nurture numerous farmer partnerships in many value chains. These cover multiple crops and geographies. Leveraging these partnerships, ITC has created a number of unique community development programmes by synergising its social sector initiatives with its business plans. ITC believes that the interdependence between its agri-based businesses and the farm sector constitutes a sustainable platform to enlarge its contribution to the Indian rural sector.

The core principles that drive these initiatives are:

- **Customise** the development model to address the diversity of rural India.
- **Enable** even the marginal farmers to access knowledge to compete on an equal footing in the market place.
- **Empower** rural communities, so that development planning and implementation are participatory.

ITC's Rural Development Initiatives Embrace Several Critical Areas

- Web-enablement of the Indian farmer to help him access relevant knowledge and services to enhance farm productivity
- Social and farm forestry to generate farm incomes in tribal hinterlands while restoring ecological balance
- Integrated watershed development to reverse land degradation and provide critical irrigation
- Economic empowerment of women to transform them into powerful agents of social change
- Primary education for the rural poor to enhance employability

ITC's e-Choupal Movement—Traditional Farmers to New-age Marketers

The immense potential of Indian agriculture is waiting to be unleashed. The endemic constraints that shackle this sector are well known—fragmented farms, weak infrastructure, numerous intermediaries, excessive dependence on the monsoon, variations between different agro-climatic zones, among many others. These pose their own challenges to improve productivity of land and quality of crops. The unfortunate result is inconsistent quality and uncompetitive prices, making it difficult for the farmer to sell his produce in the world market.

ITC's trail-blazing answer to these problems is the e-Choupal initiative; the single-largest information technology-based intervention by a corporate entity in rural India. Transforming the Indian farmer into a progressive knowledge-seeking citizen. Enriching the farmer with knowledge; elevating him to a new order of empowerment is the mandate of this approach.

Through the e-Choupal initiative, ITC aims to confer the power of expert knowledge on even the smallest individual farmer. Thus enhancing his competitiveness in the global market.

Given the low level of literacy in the rural sector, the role of the Choupal Sanchalak, the lead farmer of the village, in facilitating physical interface between the computer terminal and the farmers is central to project e-Choupal.

The Big Picture

ITC's International Business Division, one of India's largest exporters of agricultural commodities, has conceived e-Choupal as a more efficient supply chain aimed at delivering value to its customers around the world on a sustainable basis.

The e-Choupal model has been specifically designed to tackle the challenges posed by the unique features of Indian agriculture, characterized by fragmented farms, weak infrastructure and the involvement of numerous intermediaries, among others.

The Value Chain—Farm to Factory Gate

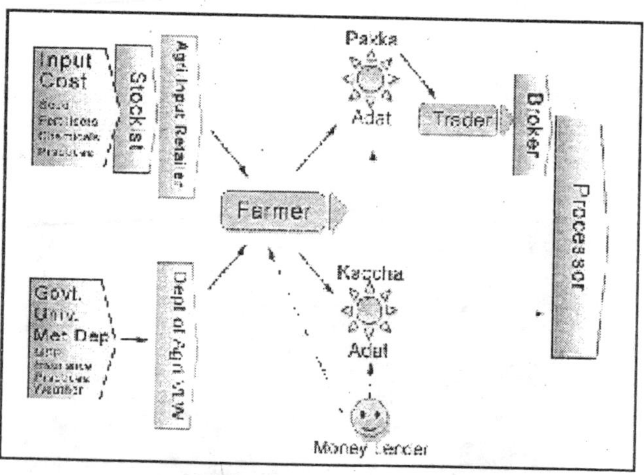

'e-Choupal' also unshackles the potential of Indian farmer who has been trapped in a vicious cycle of low risk taking ability > low investment > low productivity > weak market orientation > low value addition > low margin > low risk taking ability. This has made the farmers and Indian agribusiness sector globally uncompetitive, despite rich and abundant natural resources.

Such a market-led business model can enhance the competitiveness of Indian agriculture and trigger a virtuous cycle of higher productivity, higher incomes, enlarged capacity for farmer risk management, larger investments and higher quality and productivity. Further, a growth in rural incomes will also unleash the latent demand for industrial goods so necessary for the continued growth of the Indian economy. This will create another virtuous cycle propelling the economy into a higher growth trajectory.

The Model in Action

Appreciating the imperative of intermediaries in the Indian context, 'e-Choupal' leverages Information Technology to virtually cluster all the value chain participants, delivering the same benefits as vertical integration does in mature agricultural economies like the USA.

'e-Choupal' makes use of the physical transmission capabilities of current intermediaries – aggregation, logistics, counter-party risk and bridge financing –while disintermediating them from the chain of information flow and market signals.

With a judicious blend of click and mortar capabilities, village internet kiosks managed by farmers – called *sanchalaks* – themselves, enable the agricultural community access ready information in their local language on the weather and market prices, disseminate knowledge on scientific farm practices and risk management, facilitate the sale of farm inputs (now with embedded knowledge) and purchase farm produce from the farmers' doorsteps (decision making is now information-based).

e-Choupal delivers real-time information and customised knowledge to improve the farmer's decision-making ability, thereby better aligning farm output to market demands; securing better quality, productivity and improved price discovery. The aggregation of the demand for farm inputs from individual farmers gives them access to high quality inputs from established and reputed manufacturers at fair prices. As a direct marketing channel, virtually linked to the 'mandi' system for price discovery, 'e-Choupal' eliminates wasteful intermediation and multiple handling. Thereby it significantly reduces transaction costs.

'e-Choupal' ensures world-class quality in delivering all these goods and services through several product/service specific partnerships with the leaders in the respective fields, in addition to ITC's own expertise.

While the farmers benefit through enhanced farm productivity and higher farm gate prices, ITC benefits from the lower net cost of procurement (despite offering better prices to the farmer) having eliminated costs in the supply chain that do not add value.

A digital transformation

ITC began the silent revolution of rural India with soya growers in the villages of Madhya Pradesh.

For the first time, the stereotype image of the farmer on his bullock cart made way for the e-farmer, browsing the e-Choupal website. Farmers now log on to the site through Internet kiosks in their villages to order high quality agri-inputs, get information on best farming practices, prevailing market prices for their crops at home and abroad and the weather forecast – all in the local language. In the very first full season of e-Choupal operations in Madhya Pradesh, soya farmers sold nearly 50,000 tons of their produce through the soyachoupal Internet platform, which has doubled since then. The result marks the beginning of a transparent and

ITC's mobile vans take the message of e-Choupal to new villages. Thereafter, virtual helpdesks enable the farmer to find solutions to his problems through online interactions. ITC has set up VSAT links to overcome connectivity problems.

cost-effective marketing channel. Bringing prosperity to the farmers' doorstep.

Smart Cards enable farmer identification to provide customized information on the soyachoupal.com website. Online transactions are captured to reward farmers for volume and value of usage.

Linking Farmers to Remunerative Markets

Farmers grow wheat across several agro-climatic zones, producing grains of varying grades. Though these grades had the potential to meet diverse consumer preferences, the benefit never trickled down to the farmers, because all varieties were aggregated as one average quality in the mandis. Enter ITC's e-Choupal intervention. The-Choupal site is now helping the farmers discover the best

Choupal Sanchalaks take an oath to use the e-Choupal for the benefit of the farming community.

price for their quality at the village itself. The site also provides farmers with specialised knowledge for customising their produce to the right consumer segments. The new storage and handling system preserves the identity of different varieties right through the 'farm-gate to dinner-plate' supply chain. Encouraging the farmers to raise their quality standards and attract higher prices.

Managing Risks through Technology

The whats and ifs in the aqua farmers' life posed daunting odds. They were haunted by the nightmare of contaminated soil, wrong levels of salinity in the water or the killer White Spot virus, any of which could wipe out an entire shrimp crop. Until ITC's aquachoupal site provided them the support and the know-how to cope with and manage such risks. Information on the aquachoupal site equips them with comprehensive know-how to keep abreast of food safety norms to compete in the international market. Information includes parameters for antibiotic usage, hygienic washing, sanitised dressing and air-tight packing. All these factors help to neutralise the risks involved in aqua farming and making it economically much more attractive and benefiting hundreds of aqua farmers.

In the high-risk business of shrimp farming, the wealth of information provided by aquachoupal.com has proved a great boon for farmers in Andhra Pradesh. This success has inspired ITC to plan its expansion to West Bengal.

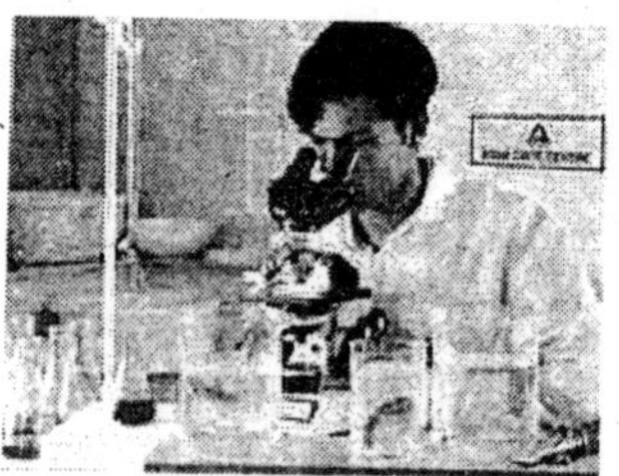

ITC's Aqua Care Centre in Kakinada, Andhra Pradesh, has revolutionised the concept of shrimp seed testing. Its sophisticated laboratory detects the deadly White Spot virus in the shrimp seed and advises farmers on appropriate remedial action.

A Dependable Knowledge Partner

A dependable knowledge partner Coffee planters in India have for years been tossed between the highs and lows of the international coffee market. The information needed to manage risks in the volatile global coffee market, price updates and prevalent trends in coffee trading were just not available to them. The launch of ITC's plantersnet.com has equipped India's coffee planters with appropriate knowledge base and risk management tools. The site arms them with the latest prices posted on commodity exchanges like CSCE in New York and LIFFE in London.

Plantersnet.com has become popular among coffee growers as an effective platform for global trade.

In addition to assisting with knowledge management through the website, ITC provides on-ground inputs to coffee planters on best practices, grading standards, quality policy etc.

The Status of Execution

Launched in June 2000, 'e-Choupal', has already become the largest initiative among all internet-based interventions in rural India. 'e-Choupal' services today reach out to more than a million farmers growing a range of crops - soyabean, coffee, wheat, rice, pulses, shrimp - in over 18,000 villages through 3000 kiosks across five states (Madhya Pradesh, Karnataka, Andhra Pradesh, Uttar Pradesh and Maharashtra).

The problems encountered while setting up and managing these 'e-Choupals' are primarily of infrastructural inadequacies, including power supply, telecom connectivity and bandwidth, apart from the challenge of imparting skills to the first time internet users in remote and inaccessible areas of rural India.

Several alternative and innovative solutions—some of them expensive—are being deployed to overcome these challenges e.g. Power back-up through batteries charged by Solar panels, upgrading BSNL exchanges with RNS kits, installation of VSAT equipment, mobile Choupals, local caching of static content on website to stream in the dynamic content more efficiently, 24x7 helpdesk etc. ITC has taken care to involve farmers in the designing and management of the entire 'e-Choupal' initiative. The active participation of farmers in this rural initiative has created a sense of ownership in the project among the farmers. . They see the 'e-Choupal' as the new age cooperative for all practical purposes. This enthusiastic response from farmers has encouraged ITC to plan for the extension of the 'e-Choupal' initiative to altogether 15 states across India over the next few years. On the anvil are plans to channelise services related to micro-credit, insurance, health and education through the same 'e-Choupal' infrastructure.

"A quiet digital revolution is reshaping the lives of farmers in remote Indian villages. In these villages, farmers grow soyabeans, wheat and coffee in small plots of land, as they have for thousands of years. A typical village has no reliable electricity and has antiquated telephone lines. The farmers are largely illiterate and have never seen a computer. But farmers in these villages are conducting e-business through an initiative called e-Choupal, created by ITC, one of India's largest consumer product and agribusiness companies."

e-Choupal—at a Glance

Milestones

- Commencement of initiative: 2000
- States covered: 5
- Villages covered: 18,000
- e-Choupal installations: 3,000
- Empowered e-farmers: 1.8 million

Agenda for the Next Decade

- States to be covered: 15
- Villages to be covered: 1,00,000
- e-Choupal installations: 20,000
- e-farmers to be empowered: 10 million

8

Information and Communication Technology (ICT) as an Alternative Extension Approach: A Critical Analysis of ICT Projects in India

Shaik. N. Meera, Anita Jhamtani and DUM Rao

ABSTRACT

In the changing agriculture scenario the use of Information Communication Technologies (ICT) in Agricultural Extension systems has become significant. There are many possibilities for potential application of the Information Communication Technology in Agricultural Extension. For developing strategy to design a new extension approach, the existing isolated ICT projects need to be critically studied on some parameters. Such parameters include process impacts, factors related to farmers influencing the success of ICT projects and effectiveness of functionaries involved in such projects. The present study was carried out to investigate into these researchable issues. Three ICT projects - Gyandoot Project, Warna Wired Project and iKisan Project were selected for the study. Analyses of 'process impacts' revealed that these projects have made effective contribution and were adequately equipped in carrying out the activities. It was found that majority of the users were small and marginal farmers. The study outlined the important information needs for these three different regions of India. It is also revealed that irrespective of computer illiteracy, ICT could be harnessed effectively

with the human interfaces in the form of functionaries at kiosks. The effectiveness of the functionaries was found to be moderate to high. Based on the study it is recommended that, ICT should be incorporated in extension systems to back-up resource poor farmers. Future initiatives should have effective functionaries who have high orientation towards ICT extension, faith in people and basic agricultural education. It is also recommended that similar research have to be carried out extensively to evolve a platform to design strategies in developing world to harness ICT in agricultural development.

Introduction

The Information and Communication Technologies (ICTs) can create new opportunities to bridge the gap between information haves and information have-nots in the developing countries (Government of India (GOI), 2000). The task force on 'India as Knowledge Superpower (Government of India 2001) emphasized the need to harness ICTs for societal transformation. The agriculturally rich developing countries like India cannot overlook agriculture in such transformation. The emerging ICTs have significant role to perform in agricultural development, as was evident from the interdisciplinary dialogue on Information Technology. Reaching the Unreached (Swaminathan, 1993). Use of ICTs in agricultural and rural development was explained by Richardson (1996). There are many possibilities of incorporation of ICTs in agricultural extension, for the overall agricultural and rural development (Zijp, 1994).

In spite of the tremendous potentials of ICTs in agricultural development, the developing countries have not adapted a sound strategy in harnessing ICTs in Agricultural Extension systems. Some isolated ICT projects were initiated in the developing world to harness ICT in agricultural development on location specific basis. But there is not enough effort to formulate a strategy for incorporating ICTs in agricultural extension systems. For evolving such a new approach, agricultural extension research have to be carried out to get the insights into working of the isolated ICT projects. There are three basic areas that are to be focussed for evolving an alternate extension approach. (a) Assessing the process impacts of the existing ICT projects of India in order to see the feasibility of ICTs in Agricultural Extension systems (b) User analysis of the Projects so as to know who will be the prospective users of the ICT Extension systems. Analysis of the farmers' information needs. (c) Assessing the effectiveness of the functionaries of the existing ICT projects so as to design the man power requirements of the future alternate ICT Extension Approach.

So far, not much effort has been put forth by the Extension researchers to critically analyse the existing ICT projects in India, in order to draw lessons and to develop insights for designing an alternate ICT Extension Approach.

In this backdrop, for the first time in India, the present research was conducted with the following objectives:

(a) To evaluate the process impacts of selected agricultural ICT projects of India.

(b) To study the factors related to farmers influencing success of such ICT projects.

(c) To assess the effectiveness of grass root level ICT functionaries.

Methodology

For conducting the study three ICT projects, out of six major ICT projects in India, were selected. The selection of the projects was purposive in nature. The ICT projects thus selected were:

Gyandoot ICT Project in Madhya Pradesh State
Warna Wired village ICT Project in Maharastra State
IKisan ICT Project in Andhra Pradesh State

These three ICT projects were purposively selected for the following reasons. (1) To generalize the findings of any research, the research is supposed to be carried out in heterogeneous conditions. In India, the ICT initiatives for agricultural development can be broadly classified into three categories. They are Govt. initiatives, Cooperative initiatives and Private initiatives. Hence, for conducting the study, *Gyandoot* ICT project as Govt. initiated, *Warna* Wired Village ICT project as Cooperative initiated and *IKisan* ICT project as Private initiated were selected. Another reason was geographical and agro-climatic conditions under which they were operating. *Gyandoot* project was started in tribal dominated drought prone area. *Warna* project being operated in agriculturally prosperous area and *Ikisan* in agriculturally commercialised area.

All the farmers using the ICT services (irrespective of their level of usage) and the functionaries of these three projects constituted the population of the study. Out of that, 80 farmers and 60 functionaries from each project were randomly selected. Thus, the total sample comprised 240 farmers and 180 functionaries in the selected ICT projects.

Data Collection Tools

In order to collect relevant information, two types of structured interview schedules were developed. One schedule was meant for collecting data from the farmers and the second from the functionaries.

Variables and Measurement

Various variables to fit in line with the objectives were considered for this purpose. The variables and their measurement are mentioned below.

To evaluate the process impacts of ICT projects, Scott Mc Connel's Model (2001) was used for the first time in India. The model comprises indicators for the process impacts of the ICT projects. In the present study, several variables were studied such as number of farmers accessing services, gender equity, user equity, technical aspects, etc. To study factors related to farmers influencing the success of ICT projects, the variables such as frequency of use, computer literacy, landholding of farmers, information needs of farmers, goal commitment, age etc. were studied. To assess effectiveness of functionaries, the variables such as Effectiveness, Orientation to ICT extension, Training received, Faith in people, Professional qualifications in Computer were considered. In the present study, effectiveness of the functionary with respect to his job refers to the extent to which he considers himself as competent individual to carry out the job. For measuring effectiveness of functionaries the scale developed by Singh and Kumari (1997) was used with some modifications so as to fit into the context of ICT. The modified scale comprised 10 statements, each statement carrying five levels of responses with the scores ranging from one to five. The total score thus ranged from 10 to 50. For measuring orientation to ICT extension, orientation scale developed by Anil Kumar (1997) was used with some modifications. To assess faith in people scale developed by Mehta (1989) was used. For other variable, schedules were developed.

Method of Data Collection

Selected respondents (farmers, functionaries) was personally interviewed by the researcher using the two structured schedules. This was supported by the personal observations and informal talks. For collecting data related to Process impacts, the official records of projects were also consulted.

Research Design and Statistical Analysis

Survey research design was used for the study. The data were entered and analysed using SPSS (Statistical Package for Social Sciences). Descriptive statistics (frequency tables, simple percentage, standard deviation, mean) and correlations were worked out for analysing and interpreting the data.

Results and Discussion

This section of the paper comprised the results of the study and the discussion that flows from the results. However, because of the space limitation, only major findings of the study are mentioned here. The results are presented in three sub-section based on the objectives of the study.

I. Process Impacts of the ICT Projects

Table 8.1: Process Impacts of ICT Projects Based on Mc Connell's Model

Indicators	*Gyandoot*	*Warna*	*Ikisan*
Number of Kiosks	39	56	18
Number of farmers accessing ICT Services per week	2340	4250	810
User Equity Male to Female Ratio	85.5:14.5	86:14	98:2
Ratio between the Users from Below poverty line and Above poverty line	34:66	28:72	16:84
Degree to which ICT used as information gathering tool	Very Rare (Once a month)	Rare (Once a fortnight)	Rare (Once a day)
As information sharing tool	Very frequent (Once a day)	Very frequent (Once a day)	Very frequent (Once a day)
Technical Evaluation Data base Management System	Application Server Type	Application Server & Internet Server	Application Server & Internet Server
Local language software	Hindi	Marathi	Telugu
Network connectivity	Fiber Optics, wireless in local loop	Fiber Optics, Wireless local loop	Fiber Optics, leased line
Scalability	21 to 39	51 to 56	9 to 18

The major findings regarding the process impacts of the three ICT projects based on the Mc Connel's Model (2001) are depicted in Table 8.1. Perusal of Table 8.1 reveals that, the number of farmers accessing the ICT services was found to be encouraging in all the three project areas. The user equity of the projects services was determined on two dimensions. One was gender equality among the users and the other one was economic equality (based on poverty line). It was revealed that there was no user equity as far as gender was concerned, in all the project areas. For effectively harnessing the benefits of ICT services, the initiatives in developing countries should try to encourage participation of women in such initiatives. Interestingly, based on the poverty line (economic status), the user equity was found to be very high. From these ratios, it can be concluded, even though all the three projects were successful in reaching out to farmers below poverty line, the extent of penetration was relatively more in case of Government Project (*Gyandoo*t) and Cooperative Project (*Warna* Project) compared to a Private Sector project (*Ikisan* project).

Mc Connell's indicators revealed that the degree to which ICT is used, as an information-gathering tool was ranging from very rare to rare. Whereas ICT use as an information-sharing tool was 'very frequent'. The technical evaluation as per the Mc Connell's indicators revealed that the Data base Management System (DBMS), Connective aspects and the use of local language (vernacular) software were adequate and made the projects efficient. Interestingly, the scalability aspects were found to be encouraging in all the three projects. In case of *Gyandoot*, the number of Kiosks was increased from 21 to 39, in *Warna* project 51 to 56 and in *Ikisan* the number rose from 9 to 18. This scalability revealed that because of the process impacts made by these projects more and more number of Kiosks was being established by the projects. It was found that, there were some problems with regard to the economic sustainability of the Kiosk operator in case of *Gyandoot* and iKisan. But, since the focus of the study was implying the use of ICTs in Extension system, which is funded by the government, there will not be any such problem.

II. Farmers' Related Factors

The important link in the whole chain of ICT networks and their applications are the ultimate beneficiaries of the ICT services i.e. farmers. Hence, the factors related to farmers are important to make any effort to incorporate ICTs in agricultural extension. Accordingly, the present study focussed on age, economic status (land holding), frequency of use of ICT services, computer literacy and information needs of the farmers.

Distribution of the respondents is presented in Table 8.2. The frequency distribution of ICT users was highly skewed towards younger age. In the total sample, about 38 per cent of farmers belonged to young age and nearly 52 per cent belonged to middle-aged group. However, the per cent of old farmers who were using ICT services was not insignificant. This is in fact, an encouraging sign. Another interesting finding, which has far reaching consequence particularly in the ICT initiatives of developing countries, is the distribution of farmer users based on landholdings. In the total sample, nearly 21 per cent of the farmer users were possessing marginal land holdings and about 47 per cent possessing small landholdings. The fact that 68 per cent of the users of ICT services were small and marginal would shatter the myth that "ICT is not for small and marginal'. However, the per cent of small and marginal farmer users was lower in case of private ICT project i.e. *Ikisan*. Hence, there is a need to incorporate ICTs in Govt. agricultural extension system to develop a strong bias towards resource poor and marginal farmers. The findings revealed that the incorporation of ICT as an alternative Extension system would have far reaching implications in reaching out to the small and marginal farmers.

Table 8.2 also reveals the frequency of use of ICT services by the farmer users. This revealed that all the users were accessing ICT services at least once in a month, if not more than that.

Another interesting finding was that only 11 per cent of the farmer users were having some knowledge about computers. Majority of the farmers (89 per cent) were completely computer illiterates. This implies that in developing countries, farmers need not be computer literate to harness the ICT in agricultural development. This reveals that ICT as an alternative Extension system would not demand for the computer literate farmers in any way.

Table 8.2: Percentage Distribution of Farmers based on Age, Landholding, Frequency of use and Computer Literacy

Variables	*Gyandoot n=80*	*Warnan n=80*	*Ikisan n=80*	*Total Sample n=240*
Age				
Young (<35 years)	45.00	35.00	32.50	37.50
Middle (35– 50 years)	45.00	62.50	47.50	51.66
Old (>50 years)	10.00	2.50	20.00	10.83
Land Holding				
Marginal (< 1 ha)	20.00	35.00	5.00	20.80
Small (1-2 ha)	50.00	42.50	50.00	46.70
Medium (2-5 ha)	25.00	22.50	42.50	29.16
Large (>5 há)	5.00	0.00	2.50	3.33
Frequency In Use				
More than twice per week	0.00	5.00	0.00	1.66
Once a week	2.50	5.00	7.50	95.00
Once a fortnight	45.00	45.00	40.00	43.33
Once a month	52.50	45.00	52.50	50.00
Computer Literacy				
Computer Literates	7.50	12.50	15.00	11.66
Computer Illiterates	92.50	87.50	85.00	88.33

Table 8.3 depicts some specific correlations found in the data analysis, which are worth mentioning in the context. Most importantly, the frequency of use of ICT services was not at all correlated significantly with the land holding of the farmers. This indicates that irrespective of the land holding ICTs were helpful to all sections of the farmers. In case of iKisan, the farmers in this region were all progressive irrespective of their age and education. Hence the correlation coefficients were not significant in iKisan.

Table 8.3: Some Specific Correlations found in Data Analysis

	Correlation Coefficient 'r' value			
Correlation between variables	Gyandoot n=80	Warnan n=80	Ikisan n=80	Total Sample n=240
Frequency of use and age	-0.685**	-0.610**	0.180	-0.500**
Frequency of use and education	0.588***	0.5212**	0.132	0.375**
Frequency of use and landholding	0.277	0.163	0.414	0.121

** Significant at 0.01 level of probability.

Current evidences suggest that, to achieve sustainability and success, ICT project for agricultural development must begin with the real needs of the local community of farmers (Richardson, 1996). Unfortunately, the task of understanding the clientele farmers and their information need had been subsided by the technological enthusiasm that is prevailing in developing countries. The information needs of the farmers with respect to their perception about importance attached are presented in Tables 8.4, 8.5 and 8.6.

In case of Gyandoot project, majority farmers perceived market information, facilitation of land record, question-answer services, information on Rural Development

Tables 8.4: Information Needs of Farmers—Gyandoot Project: Percentage distribution of Farmers based on their Perception about Information Needs

Information Needs	*Most Important*	*Important*	*Less Important*
Question Answer Services	67.50	32.50	0
Market Information	90.00	7.50	2.50
latest (best) package of practices	12.50	70.00	17.50
Disease/Pest early warning system and management	0	30.00	70.00
Weather forecasting	47.50	45.00	7.50
Information on rural development programs/subsidies	57.50	27.50	15.00
Directory and Information on Crop Insurances	5.00	55.00	40.00
General Agricultural News	2.50	67.50	30.00
Post harvest technology	10.00	70.00	20.00
Facilitation of land records/registration online	82.50	17.50	0

Programs and Weather forecasting as the most important and essential information needs. Likewise information needs on best package of practices, information on post harvest technology, agricultural news and crop insurance information were rated as somewhat important. Farmers did not perceive the remaining information needs as that essential (please see Table 8.4). In case of Warna project, majority of farmers perceived question-answer services. Accounting-Payments of Cooperative best practices and market information as the most essential and important information needs. Similarly, information needs like input prices/availability, early warning

Table 8.5: Information Needs of Farmers—Warna: Percentage distribution of Farmers based on their Perception about Information Needs

Information Needs	*Most Important*	*Important*	*Less Important*
Question Answer Services	87.50	12.50	0
Market Information	50.00	45.00	5.00
Latest (best) package of practices	67.50	32.50	0
Disease/Pest early warning system and management	47.50	52.50	0
Input prices and availability	17.50	57.50	25.00
Weather forecasting	17.50	37.50	45.00
Information on rural development programs/subsidies	15.00	52.50	32.50
General Agricultural News	12.50	47.50	40.00
Facilitation of land records/registration online	30.00	30.00	40.00
Accounting and easy payment services	75.00	25.00	0

systems, information on RD programs and general agricultural news as somewhat important information needs. Remaining information needs were not perceived as that essential. In Ikisan project area, majority farmers perceived early warning systems, information on RD programs, question-answer services, information on cropping systems and planning and best practices as the most essential and important information needs.

The findings revealed that even though there are different information needs for the farmers in different farming systems, there are some common information needs such as market information, question answer services, etc.

Table 8.6: Information Needs of Farmers—iKisan: Percentage distribution of Farmers based on their Perception about Information Needs

Information Needs	*Most Important*	*Important*	*Less Important*
Question Answer Services	60.00	40.00	0
Market Information	35.00	50.00	15.00
Latest (best) package of practices	52.50	42.50	5.00
Disease/Pest early warning system and management	65.50	32.50	2.50
Input prices and availability	32.50	40.00	27.50
Weather forecasting	27.50	60.00	12.50
Information on rural development programs/subsidies	65.00	30.00	5.00
Directory and Information on Crop Insurances	32.50	42.50	25.00
General Agricultural News	5.00	57.50	37.50
Information on Cropping systems and planning	45.00	50.00	5.00
Facilitation of land records/registration online	60.00	35.00	5.00

Hence it could be said that the alternative (ICT) extension system would require providing the information services in three levels. (1) at region level (state) level which include information such as facilitation of land records, agricultural news and insurance information, information on rural development programmes of that region. (2) At district level-input prices/availability, weather forecasting, early warning systems, information on post harvest technology, rural development schemes etc. (3) At block level-question-answer services, best package of practices etc.

III. Effectiveness of the Functionaries

Table 8.7 presents the results regarding the distribution of functionaries based on effectiveness, orientation to ICT extension, education and professional qualifications in computers. About 65 per cent of functionaries working in ICT projects showed moderate effectiveness and 18 per cent showed higher level of effectiveness. Skewness towards higher level of effectiveness is a good indicator for the organisational performance of ICT Projects. Likewise, in case of orientation towards ICT extension majority of the functionaries showed either medium or higher favourable orientation, which should be the pre-requisite for success of ICT projects.

The implications of these findings on the ICT Extension approach are quite many. The Extension workers in the new approach need to be highly effective in harnessing the ICT for agricultural development. There is a need to inculcate the orientation towards ICT extension approach among the grass-root level extension functionaries.

Table 8.7 also provides frequency distribution of functionaries based on education and professional qualifications in computers. This revealed that, the human interfaces between ICT networks and farmers need not be having higher level of education and professional qualification in computers. It could be said that new ICT Extension Approach would not require extension workers to have professional qualification in computers. Simple working knowledge of computers is sufficient for the extension functionaries to effectively harness the ICT an alternative Extension Approach.

To explore into the factors associated with effectiveness of functionaries, correlation analysis was

Table 8.7: Percentage distribution of functionaries based on Effectiveness, Orientation to ICT Extension, Education and Professional Qualification

Variable	*Gyandoot* *n=60*	*Warnan* *n=60*	*iKisann* *n=60*	*Total* *n=180*
Effectiveness				
Low	13.30	23.30	20.00	17.70
Medium	73.30	50.00	70.00	65.50
High	13.30	26.60	10.00	16.66
Orientation to ICT Extension				
Less	34.40	16.60	16.60	11.10
Medium	50.00	56.60	70.00	75.50
High	16.66	26.66	13.33	13.33
Education				
Middle School	3.33	0	0	1.11
High School	50.00	23.33	6.66	26.66
Graduate	33.33	50.00	36.66	40.00
Professional Degree				
(Ag. B.Sc.)	13.33	26.66	56.66	32.22
Professional Qualification in Computers				
No qualification	16.66	26.66	30.00	24.44
Diploma	56.66	73.33	53.55	61.11
Degree	26.66	0	16.66	14.44

carried out. Table 8.8 provides results regarding the same. In total sample, education, orientation towards ICT extension and faith in people are significantly and positively correlated with effectiveness, at 0.01 level of significance. Where as professional qualification (B. Sc Agriculture) was correlated significantly at 0.05 level of significance. These factors if kept in mind may help design a manpower requirement of the future ICT Extension Approach.

Table 8.8: Correlation Coefficients of effectiveness of Functionaries with other Variables

Variables	*'r' value* Gyandoot	*Warna*	*iKisan*	*Total Sample*
Education	0.648**	0.352	0.420*	0.515**
Professional Qualification	-0.423**	0.086	0.130	0.237**
Orientation to ICT Extension	0.644**	0.523	0.665**	0.600**
Faith in People	0.738**	0.614	0.637**	0.694**

* Significant at 0.05 level of probability.

** Significant at 0.01 level of probability.

Conclusions and Recommendations

The study revealed encouraging signs of harnessing ICT as an alternate approach in India in particular and developing countries in general. Based on the findings related to process impacts, factors related to farmers influencing success of ICT initiatives and the effectiveness of functionaries, following conclusions are made:

1. Since the process impacts of ICT projects of ICT were encouraging on different indicators of Mc Connell Model, it is recommended that ICTs be incorporated in Agricultural Extension for all endeavours of agricultural development. The findings revealed that the incorporation of ICT as an alternative Extension system is a reality and will have far reaching implications in reaching out to the small and marginal farmers.

2. Generating awareness about availability of ICT services among young and middle age farmers, first step to be considered soon after initiating ICT projects/ ICT extension approach in future. The old farmers should be brought into the chain of ICT networks in the later stages.

3. Since all the categories of farmers were using ICT services irrespective of land holding, it can be said that all the farmers can be reached using ICT. However, in case of Government initiative and Cooperative initiatives, number of small and marginal farmers were using ICT services. There is a need to incorporate ICT in Government extension system to develop a strong bias towards resource poor farmers.

4. The findings revealed that even though majority of farmers were computer illiterate, if the human interfaces (Kiosk operators/functionaries) were established, ICTs could be a success. In developing countries, hence, all the future ICT endeavours should focus on how to build as sustainable human interfaces. Hence, the future alternate extension system should develop a strong interface between the ICT and the farmers.

5. The ICT project must take into account the information needs of the farmers. In the less endowed areas, agriculturally commercialised areas and in the areas where cooperative set-ups exist, the information needs of farmers, as revealed by the study, must be integrated in respective ICT projects. Same can be done in the future ICT extension approaches also.

6. The findings revealed that even though there are different information needs for the farmers in different farming systems, there are some common information needs such as market information, question answer services, etc. Hence it could be said that the alternative (ICT) extension system would require providing the information services in three levels. (1) At region level (state level) that includes information such as facilitation of land records, agricultural news and insurance information, information on rural development programmes of that region. (2) At district level-input prices/availability, weather forecasting, early warning systems,

information on post harvest technology, rural development schemes etc. (3) At block level-question-answer services, best package of practices etc.

7. The future ICT project/initiatives should have highly effective functionaries who act as interfaces. While manning such projects factors like their orientation towards ICT extension, faith in people, education and professional qualifications will have to be considered. The Extension workers in the new approach need to be highly effective in harnessing the ICT for agricultural development. There is a need to inculcate the orientation towards ICT extension approach among the grass-root level extension functionaries.

8. Based on the success of ICT projects in reaching farmers as revealed in the study, it is recommended that Governments have to re-orient their agricultural policies so that a full-fledged strategy is developed to harness ICT at all levels.

8. The Extension system in India has to gear up in evolving a strong ICT-Extension strategy so as to incorporate the ICT as an Agricultural Extension system.

It is also recommended that based on the methodologies adopted in this study, similar researches should be carried out in all the project areas of India and other developing counties. This will help in developing insights and draw lessons. A series of similar researches should be carried out to develop a new research paradigm viz. "ICTs in agricultural development". Such a new paradigm shall help designing a perfect ICT-Extension systems, which will enable India realise the concept of Knowledge Society. In a knowledge Society everything is possible including Sustainable Agricultural Development.

References

Anil Kumar (1997). Communication effectiveness of agricultural officers in Kerala, India. A Psycho—personal analysis. Unpublished Ph.D Thesis, IARI, New Delhi.

Government of India, (2001). Ministry of Information Technology. Report of Task Force on India as knowledge Super power. Planning Commission, New Delhi.

Government of India, (2000). Ministry of Information Technology. "Working group on information technology for masses". Background Report "On line http://infomasses. Nic.in/ vsitformasses/page1 htm.

Mc Connell, Scott (2001). Connecting with the Unconnected: Proposing an Evaluation of impacts of internet._Mc Connell International (on line http:/mcconnellinternatinal.com/ evaluatuin.htm.

Mehta, A (1989). In: Singh, L. (1998) Dynamics of Participation in Doon Valley Watershed Project. Unpublished Ph.D. Thesis, IARI, New Delhi.

Richardson, D. (1996). The Internet and rural development recommendations for strategy and activity. Final Report. Sustainable Development Department of FAO, http:// www.fao.org/sd-dimensions/

Singh, A.P and Kumari Patiraj (1997). In: Third Hand Book of Psychological and Social Instruments. Vol.2 ed. O.M. Pestonjee. Concept Publishing Company, New Delhi, pp. 294-296.

Swaminathan, M.S. (1993). (Ed). *Information Technology: Reaching the Unreached* Chennai: Macmillan. India.

Zijp, W. (1994). Improving the Transfer and Use of Agricultural Information—A Guide to Information Technology. World Bank, Washington DC.

9

Implementation Strategy in World Bank Funded "Diversified Agriculture Support Project (DASP)" in State of Uttar Pradesh with Emphasis on Multiple Approach of Extension for Dissemination of Technology

Surya Pratap Singh and Mukesh Gautam

ABSTRACT

This paper describes in detail the various extension methodologies adopted in World Bank funded "Diversified Agriculture Support Project" in selected 157 blocks of 32 districts of Uttar Pradesh. Launched in 1998 with the objective of increasing agriculture productivity through support for UP's diversified production systems, encouragement to privatisation in agriculture and improvement in rural infrastructure, the project is now entering in last year of its 5 year phase. In the last four years the areas of project activities have centered around improving food security with focus on diversified agriculture growth and increased local participation, rural restructuring and public reforms related to decentralisation, policy reforms and cost recovery. Enhancing institutional strength through capacity building of all stakeholders (farmers, NGOs, Groups, Line agencies, Panchayati Raj institutions) has been one of the major activities in addition to improving rural infrastructure and conserving biodiversity. Need based development and dissemination of new technology and spread of existing technology among rural masses have immensely benefited with the project towards achieving the target of desired productivity. Several methodologies have been field tested to disseminate the technology. This paper also deals with each one of them and their relative importance under different situation and field conditions.

Introduction

This paper describes in detail the various extension methodologies adopted in World Bank funded "Diversified Agriculture Support Project" in selected 157 blocks of 32 districts of Uttar Pradesh. Launched in 1998 with the objective of increasing agriculture productivity through support for UP's diversified production systems, encouragement to privatisation in agriculture and improvement in rural infrastructure, the project is now entering in last year its 5 year phase. In the last 4 years, the areas of project activities have centered around improving food security with focus on diversified agriculture growth and increased local participation, rural restructuring and public reforms related to decentralisation, policy reforms and cost recovery etc. Enhancing institutional strength through capacity building of all stakeholders (farmers, NGOs, Groups, Line agencies, Panchayati Raj institutions) has been one of the major activities in addition to improving rural infrastructure and conserving biodiversity. Need based development and dissemination of new technology and spread of existing technology among rural masses have, immensely, benefited the project towards achieving the target of increasing desired productivity. Several methodologies have been field tested to disseminate the technology. Present paper presents each one of them and their relative importance under different situation and field conditions.

Implementation Approach to Diversification and Intensification

Multifarious approach cover basic needs of farming community was adopted. Farmer's participation was ensured in need identification, development ofstrategicplans,resourseallocation implementation and monitoring.

Implementation Strategy

1. **ATMA**
2. **Farming System Approach**
3. **Participatory Planning and Implementation**
4. **Implementation through groups**
5. **Capacity building of stakeholders**
6. **Eco-friendly sustainable farming system**
7. **Privatization**
8. **CARP for technology**
9. **Multipie approach for extension**

The basic strategy of increasing agriculture productivity has been through diversification of paddy-wheat cropping rotation to more remunerative horticultural crops and intensification of agriculture through adoption of new technologies such as use of balanced fertilizers, (IPNM), adoption of Integrated Pest Management (IPM), more and more use of improved seeds of high yielding varieties, adoption of improved practices of organic composting such as NADEP, Vermi Composting and Cow Pet Pit (CPP) etc. The adoption of above-mentioned technologies was made possible through demonstrations at one or two places and its dissemination to other areas through organisation of field days at the demonstration site.

1. *Agricultural Technology Management Agency (ATMA)*

For centralized programme planning and implementation, a new agency called the 'Agriculture Technology Management Agency' (ATMA) has been created. ATMA, an autonomous body, can receive allocate and spend government funds. District level heads of all government extension departments make up ATMA Management Committee. ATMA, governing Board is composed of cross section of stakeholders including representatives of farmers and, group leaders. Board approves all block level and district level plans and approves the funds under the project. This concept has been implemented in all the 32 DASP supported districts.

2. *Adoption of Farming System Approach*

Looking at the farmers requirement for their livelihood support system, demand driven and need based interventions were made in the field of agriculture, horticulture, dairy, animal husbandry etc. Based on the needs of farming community village action plans were developed and accordingly demonstrations in agriculture, horticulture, animal husbandry etc. were planned. These included demonstrations in Integrated Plant and Nutrient Management (IPNM), Integrated Pest Management (IPM) in agriculture, and varietal demonstration in horticulture. Specific technologies such as low tunnel polyhouse for vegetable nurseries, staking in tomato or machhan cultivation for cucurbits were also demonstrated to the farmers depending upon the requirement of the area and crops grown in the area. Vaccination drives, A.I. Programme, demonstrations in nutrition management for milch animals were launched on the area specific needs of the farmers.

Adoption of Farming System

Agriculture
-IPNM
-IPM
-Seed Producer
-Organic Farming
-NADEP
-Vermi
-CPP
Horticulture
-Varietal demonstration
-specific technologies
-Machhan
-Stake
Animal Husbandry
-Vaccination
-A.I.
-Paravet services
-Nutrition Management
Dairy
-Clean milk production
-Marketing

3. ***Participatory Planning and Implementation***

In the last four years more than 6000 grama panchayats have been covered and in all the villages. Farmers have been consulted in need identification, problem solving (developing village action plan), laying of demonstrations and lastly, in the dissemination of technology to other farmers. Women are also equal partners in the process. For ensuring farmers participation in the project activities, NGOs have been actively involved from the very beginning. NGOs have been helping in creating awareness among rural masses.

Farmers' Participation

- NGOs support
- Need Identification
- Village Action Plan
- Group formation
- Demonstration
- Dissemination

4. *Implementation through Groups*

More than 19000 male and female farmers groups have been successfully formed in the project area. These groups have been used as a vehicle for carrying the technology to the farmers field. Initially developed as saving groups, these farmers groups are converting themselves into Farmers Interest Groups(FIGs) after acquiring required skills by attending capacity building programme and demonstrations of the technologies in their field.

Implementation through Groups

- **Formation of FIG/SHG**
- **Regular saving**
- **Common activity for the group**
- **Capacity building**
- **Demonstration of the technology**
- **Linkages with banks**
- **Input arrangement linkages with market**

The common activities of FIGs have been in the areas of seed production, dairy, vegetable production through organic farming, vermi composting, bee keeping, mushroom cultivation etc.

5. *Capacity Building of Stakeholders*

Once the group identifies its area of interest or area specific need emerges through Village Action Plan, groups leaders are imparted training in the identified subjects. Need based field demonstrations are also executed. Till date 37710 demonstrations in IPNM, 901 in IPM, 8203 varietal demonstrations in horticulture etc have been carried out, taking required technologies to the field. Exchange visit are also organized where farmers come in contact with others farmers and share their experiences. Farmers are regularly exposed to new technology through exposure visits to State Agriculture

Capacity Building

- **Specific training to Groups Leaders**
- **Skill upgradation through demonstrations**
- **Exposure/Exchange Visits**
- **Field days**
- **Use of electronic media (AIR/TV)**
- **Involvement of SAUs in training**
- **Linkages with SAUs/Line Agency**

Universities (SAUs), farmer's fairs and agri exhibitions etc. SAUs have made it a practice to have a follow-up with the farmers. Farmers-scientists interaction programme are organized in each cropping season.

6. Eco-friendly Sustainable Farming System

Eco-friendly Technology

- Balanced use of fertilizers
- More use of organic compost
- Green Manuring
- Integrated Pest Management
- Use of Bio-fertilizers
- Clean milk production
- Oxitocin free milk
- Drive against banned pesticides

Stress on balanced use of fertilizers after proper soil testing and demonstration of IPNM technology has immensely helped in increasing production. Construction of 7974 NADEP pits for organic composting, 4227 vermi compost pits, 10,916 CPP with little or no help from the project indicate farmer's strong inclination towards more and more use of organic composts. To give a boost to the use of eco-friendly technologies, it has been decided to demonstrate green manuring through 6000 demonstrations in Jayad 2003. In each district one village has been selected where department of agriculture and horticulture have jointly covered the entire village with Integrated Pest Management practices for every crop grown in the village. These villages are fast becoming model villages for areas in the surroundings. Use of bio-pesticides has gone up several folds. The department in clean milk production trained more than 60,000 dairy farmers. Awareness campaigns are being continuously launched against use of oxictocin in milch cattles and use of banned pesticides in crops, especially vegetables.

7. Encouragement to Privatisation in Agriculture Privatisation

In two districts, extension services were privatized and responsibilities of carrying out all departmental activities were handed over to respective NGOs working in the district. Response has been mix.

445 vegetable nursery growers were trained and developed to sell vegetable seeds to fellow villagers. Some of the NGOs were trained in bio-pesticides production. They are now selling bio-agents (trichoderma) to the farmers. Farmer Field Schools have been equipped to carryout soil testing for NPK on cost basis. Most successful example has been in the area of animal health services. More than 1100 farmers were trained as paravet workers to handle some of important animal health services such as vaccination, A.I. etc in remote areas where regular veterinary department workers have not been able to render their services.

Privatization

- Privatization of extension
- Paravets
- Capacity building of entrepreneur
- SHGs developments towards agri-gusniness
- Soil testing

8. *Competitive Agricultural Research Programme: Inclusion of recommendation in package of practices*

In the initial phase of the project, demand driven need based Strategic Research and Extension Plans were developed in all the 32 districts in consultation with local farming community. In the process researchable issues from different agro-eco situation were also identified by UP Council Agriculture Research. Proposals were screened and awarded to SAUs/NGOs/KVKs/Individuals whose recommendations or findings are regularly presented and finally included in the package of practices. Validation trials for the identified indigenous technical know-how (ITKs) have also been conducted and validated technologies (64 in number) have been recommended for dissemination.

CARP for Technology Development

- Identification of area specific issues
- Screening
- Competitive bidding by SAUs/NGOs/KVKs individuals
- Award of work
- Presentation of results
- Inclusion of

9. *Multiple Approach for Extension*

Sincere and systematic efforts of last three years are bearing fruits and impact is quite visible in the field. This has been possible due to multiple effect of various extension methodologies adopted in the field. Some of three are being presented in the following text.

(i) Strategic Research and Extension Plans (SREP)

Area specific extension issues were identified in each district and Annual Action Plan were developed by agriculture, horticulture, dairy and animal husbandry on the basis of SREPs, developed in consultation with farmers. Accordingly, interventions in the form of demonstrations/workshops/campaigns were executed with the help of NGOs and concerned line agencies. Some of the major extension issues identified through SREP are mentioned here. (a) imbalanced use of fertilizers (b) low use of organic compost (c) non availability of quality seeds (d) use of banned pesticides in several parts (e) over use of chemical pesticides, especially in vegetables (f) lack of improved breed conservation facilities (g) lack of animal health services in remote villages (h) production of unhygienic milk (i) lack of infrastructural facilities (j) lack of knowledge about improved post harvest techniques.

Multiple Extension Approach	
(i)	**Strategic Research & Extension Plan**
(ii)	**(ii)Adoption of farming system approach**
(iii)	**Group approach**
(iv)	**Use of folk media**
(v)	**Development of suitable literature for farmers**
(vi)	**Farmers oriented magazine**
(vii)	**Use of electronic media**
(viii)	**Involvement of rural students**
(ix)	**Appreciation to best farmers**
(x)	**Wall painting**
(xi)	**Frmer led extension (Farmers Field School)**
(xii)	**Field Days**
(xiii)	**Exposure/Exchange visits**
(xiv)	**Use of information technology**
(xv)	**One to one communication**

(ii) Adoption of Farming System Approach

As already indicated, integrated extension approach to issues related with the farmers' need in agriculture, horticulture, dairy, animal husbandry was adopted in the project. This has benefited small farmers immensely as their livelihood support not only comes from agriculture but it is appreciably, supported by animal husbandry, dairy, vegetable cultivation also.

(iii) Group Approach for Extension

Demonstration of new technology in each component has been done through the groups developed in the project. This has helped the group in transforming themselves in Farmers Interest Groups. Group leaders are trained in the specific technologies and from group leaders the technology is transferred to the remaining members of the group and then to the entire village. Field days are also organized at site and group members help in dissemination of the technology. Group approach has yielded in the formation of activity specific FIGs. At present 4442 groups are using organic compost produced through NADEP/CPP/Vermi Compost. In 2876 groups, members have started producing seed for the entire village communtiy. Similarly, groups are engaged in many other beekeeping, vegetable production and dairy activities etc. Banks also find it easier to lend credit to the groups than to the individuals.

(iv) Use of Folk Media

One of the most innovative methodologies has been the use of locally available folk media in delivering the technical message to the community. Project regularly identifies such types of culture parties, trains them and utilizes them for agriculture extension purposes. These have been very successful in generating awareness about the adverse impact of banned pesticides, overuse of pesticides, group formation, soil testing etc.

(v) Development of Suitable Literature for Farmers

To overcome the paucity of suitable technical literature in Hindi, series of technical bulletins, pamphlets, charts, monthly impact points have been published, separately, for farmers and project functionaries. These are extensively used in the training organized under the project. Publication of monthly impact points has touched 17,000 mark.

(vi) Publication of Farmers' Oriented Magazine

Project publishes quarterly newsletter in Hindi called Krishi Vividhikaran. It contains season specific technical information on various agricultural component, farmers experiences and success stories of the project. It also highlights specific issues related with the project. Circulation of this priced newsletter has touched 12,000 mark in last quarter.

(vii) Use of Electronic Media

Project has been broadcasting special programmes on diversification of agriculture on AIR from 6.50 pm to 7.20 pm on every Friday and Sunday. Other regular faature on AIR such as Krishi Charcha, Vigya aur Kisan and Kisan Ke Liye are also, regularly, utilized for this purpose. Timely input through Door Darshan (T.V.) is also provided to the farmers.

(viii) Involving Rural Students

On both the national days i.e. 15th August and 26th January, special campaigns on diversification, intensification, environment management are organized in the schools of project area. The basic idea behind this move has been to cover the future generation from sustaining the outputs achieved in the project.

(ix) Appreciation to Best Farmers

Innovative, highest yield producing farmers/groups leaders are rewarded at district and state level on 18th December of each year on birthday of Ch. Charan Singh. This has helped in generating tremendous enthusiasm for adopting new technology.

(x) Wall Writing

For passing loud and clear messages, wall writing is very common. To make the farmers aware about the banned pesticides wall writing proved to be quite effective. Farmers groupleaders, students help in putting the messages on the walls with almost negligible cost involved.

(xi) Farmer Led Extension Approach (Farmers Field School (FFS)

The most effective methodology of disseminating technology has been through Farmers Field Schools which are developed and self managed by the farmers themselves. Selected under the project, these self motivated farmers who have been continuously trained in different subjects, have networked or federated themselves on a common plate from and formed these farmers field schools. Their skills have been further refined by subject specific trainings in addition to trainings on PRA and other field oriented training tools. As on December, 2002 there are 123 farmers field schools operating in 157 blocks of 32 districts. These farmer trainers help the line agencies in carrying out field demonstrations, organizing field days etc. In all the schools soil testing kits have been provided on cost sharing basis. These schools aim to develop themselves in self-managed resource center of information on improved technology, marketing, inputs etc. For continuous upgradation of their skill, these schools have effective linkages with local KVK, SAUs and all the line agencies working in the project. Recently, members of these FFS have been linked with ATMA formed at district level. Private organisations like ITC, are coming forward to link themselves with these schools.

Farmer-led Extension

- **Subject specific training**
- **Identification/selection**
- **Federating FFS**
- **Developing Master Trainers**
- **Bye laws**
- **Dissemination of knowledge**
- **Organizing field days**
- **Resource center**
- **Linkages with:**
  -Research Agency(SAU/KVK/ZRS)
  -Line Agencies(DOH, DOA, Dairy, AH)
  -Private Organizations(ITC, NAFED)

(xii) Field Days

Organizing field days on demonstration sites of IPNM, IPM, varietal demonstration, machan cultivation, staking etc by department of agriculture, horticulture has helped in the dissemination of technology beyond farmer groups. Around 25000 ha of area diversified from traditional crops rotation to other remunerative horticultural crops till March, 2002.

(xiii) Exposure Visits

Farmers, group leaders, women field functionaries are regularly exposed to new technology through exposure visits to Farmer's fair organized by SAUs, agriculture exhibitions. Exchange visits of farmers within the project and outside project area are common feature of DASP activities.

(xiv) Use of Information Technology in Agriculture Extension

Project has encouraged use of information technology at various level. ATMA has been equipped with computer facility and have been linked to DASP head quarter through internet connections. All the line agency have been provided computers. Department of horticulture has established separate market information cell with internet facilities for the purpose of wholesale prices and arrival information about different horticultural crops. 16 regional Mandis have been strengthened under the project with the facility of computer alongwith Internet connection.

(xv) One to One Communication by Chief Executive

Project Coordinator of DASP continues to have one to one communication with all the groupleaders, trained farmers and other leading farmers through individual letters to them. This process is availed for passing many of key informations directly to the users from time to time.

Extension and Farmer Training of E.I.D Parry (I) Ltd.

Pillai, S.S

ABSTRACT

The corner stone of the modern agriculture is the transfer of scientific know-how to the farmer in an implementable format. This requires substantial training of extension officers and farmer training staff, who in turn will train the farmers, the youth and the farm workers. Therefore, no party can be taken for granted. Unfortunately we have many untrained graduates who are recruited fresh and are expected to carry out extension functions. The experience of these graduates is what scanty knowledge they have obtained in the universities. Therefore, the practical experience of the farmer remains superior and new knowledge fails to create an impact against the practical operations of the farmer in terms of financial gains. It is unfortunate that in such a situation many chemical companies and sellers of seeds gain the upper hand and many types of products be it registered or not, tested or otherwise are over rated by the companies in their performance and forced on to the farmers. There are many instances where the scientists succumb to the sales talk and do not question the validity or veracity of the sales propaganda e.g. application of 50 kg of BHC per acre for termite control in cane, application of endosulphon and other insecticide for early shoot borer in young cane etc. Once these were stopped there was no increase in insect pests on cane. In the sugar industry the millers were compensating for such materials regardless, as the mill staff were fairly unfamiliar with such products. It is with the aim of discouraging such wasteful practices, and improving the cropping standards, thereby increasing the farmers income and lifting the standard of the living of the farmer and converting subsidies into services. Parry undertook to develop a R&D wing with emphasis on a comprehensive training and extension programme along with in-depth researches. This paper gives a detailed outline on the management structure, implementation process, training and duties of extension envisaged by Parry in the development process. The achievements of EID Parry through Research, Extension and Training are presented in depth.

Introduction

The corner stone of the modern agriculture is the transfer of scientific know-how to the farmer in an implementable format. This requires substantial training of Extension officers and Farmer training staff, who in turn will train the farmers, the youth and the farm workers. Therefore no party can be taken for granted. Unfortunately we have many untrained graduates who are recruited fresh and are expected to carry out extension functions. The experience of these graduates is what scanty knowledge they have obtained in the universities. Therefore the practical experience of the farmer remains superior and new knowledge fails to create an impact against the practical operations of the farmer in terms of financial gains.

It is unfortunate that in such a situation many chemical companies and sellers of seeds gain the upper hand and many types of products be it registered or not, tested or otherwise are over rated by the companies in their performance and forced on to the farmers. There are many instances where the scientists succumb to the sales talk and do not question the validity or veracity of the sales propaganda e.g. application of 50 kg of BHC per acre for termite control in cane, application of endosulphon and other insecticide for early shoot borer in young cane etc. Once these were stopped there was no increase in insect pests on cane.

In the sugar industry the millers were compensating for such materials regardless, as the mill staff were fairly unfamiliar with such products. It is with the aim of discouraging such wasteful practices, and improving the cropping standards, thereby increasing the farmers income and lifting the standard of the living of the farmer and converting subsidies into services, Parry undertook to develop a R&D wing with emphasis on a comprehensive training and extension programme along with in-depth researches.

Management Structure

For any programme to be successful, leadership, management style, knowledge, training and information flow are vital components. Thus a structure was created

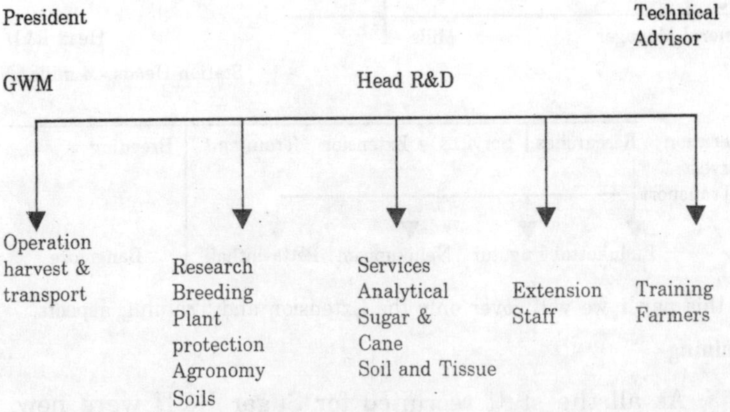

E.I.D Parry owns four mills covering an area approximately 100,000 acres. The annual cane producti

is 3.8 million tonnes in peak years with a sugar output of 3,60,000 mt.

In 1993 Parry's output from then two mills namely Nellikupam and Pugalur was 8,00,000 tonnes cane out of which 70,000 mt of sugar was produced. In 2001-2002 season the same two mill produced 20,00,000 tonnes cane which yielded 190,000 mt of sugar without any major capital undertaking.

The Cauvery sugar and Chemicals which when acquired by Parry was producing 4,00,000 tonnes cane. The production leaped to 6.2 lakh tonnes within 18 months.

A Green field factory was set up in the year 2000 by Parry at Pudukottai in an area where sugarcane was not a major crop at all. The production of sugarcane in 2002-03 was 7 lakh MT.

Implementation Programme

The field activities were divided into two sections: Crop production and Harvest and transport. Crop production includes research, extension and training.

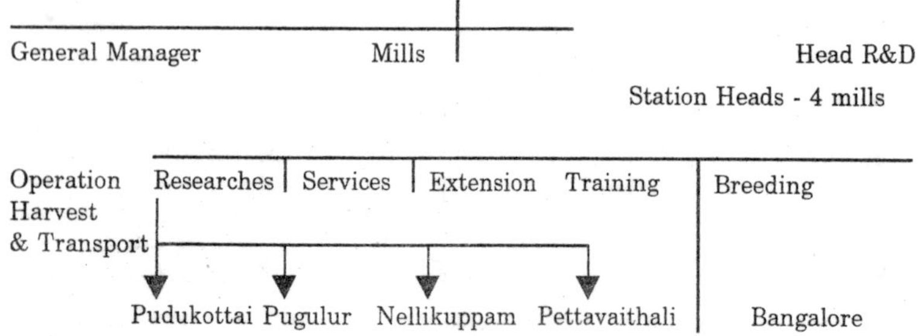

In this paper we will cover only the Extension and Training aspects.

Training

As all the staff recruited for Sugar R&D were new, they had to be given extensive training in sugarcane crop production, crop protection and crop improvement.

Technical Aspects of Cane Growing

This commences with land development which involves land clearing in case of virgin lands, and where the area is under cultivation, it has to be checked for salinity, alkalinity and nutrient levels in the soils through soil analyses. This is followed by ripping the land to a depth of 60 cm and ploughing to a depth of 45-50 cm. This is followed by either harrowing or rotovation. The furrows are formed for planting.

Sugarcane is propagated through cuttings. Seed cane is generally obtained from approved nurseries, which are free of diseases and pests. The soil analysis generally gives a fairly good indication of nutrient levels. Thus phosphate and potassium fertilizers along with a small quality of nitrogenous fertilizer is placed in the furrows and covered lightly with soil. The setts are placed in the furrows on top of the soil covering the fertilizer with the eyebuds facing sideways and covered with soil to a depth of five cm and irrigated.

Prior to planting it is desirable to form roads, irrigation channels, drains and adopt soil conservation practices. Varieties are the most important component of sugarcane cropping thus appropriate varieties is planted. Crop protection is a major exercise as sugarcane is affected by some 40 diseases, insect pests, rats and nematodes.

Table 10.1: Fertilizer Usage amount of Fertilizer to be Applied per acre

	N Urea	*P Single Super*	*K*	*Dolomite*	*Gypsum*
Total	200	80	100-200	5	10
Basal at Planting	20	60	80 - 150		
40-50days	60	-	-	-	-
90-100 days	100	20	20	-	-
180 days	20		-	-	-

Organic manure is incorporated during land preparation e.g press mud, fibre, millash, distillery waste, etc. The use of blended fertilizer is desirable.

Table 10.2: Water Usage

	Period days	*No of of irrigation once in*	*Interval Irrigation*	*No of application*	*Rate of water needs*	*Total*
Germination	0-45	45	5	9	12.5	110
Early growth	46-120	75	7	10	25.0	250
Grand Growth	Early 121-250	50	5	10	50	500
	Mid 171-250	80	7	11	50	550
	Late 251-320	70	8	9	25	225
Maturity	321-360	10	10	4	10	40
Total				53		1675

While the above indication are for optimum crop requirement, sugarcane will survive with half the precipitation, provided the distribution is good and the soils have sufficient clay and organic content.

Duties of Extension

The following activities are undertaken by extension:

- Land: Mapping the area to determine total available arable land
- Identify area for cane.
- Demarcate potential area for cane i.e. marginal and those under other crops.
- Develop fallow land for cropping.
- Determine topography & land classes and map these out.

- Land development
- Land preparation for new areas
- Clearing
- Ripping
- Ploughing, harrowing or rotovation
- Organize farm layouts
- Lay roads
- Form drains
- Soil conservation practices i.e. contour, diversion banks, grass lines, etc.
- Irrigation Channels

Planting Suitable Varieties

- Nurseries and the management of seed cane supply
- Pre-treatment of seed cane
- Basal application of nutrients
- Planting
- After planting care

Agronomic Practices

- Plant crop care
- Cultivation
- Weeding
- Ratoon: Cultivation and management
- Weed and pest control
- Gap filling

Fertilizer Usage

- Soil and leaf analysis
- Soil mapping

- Interpretation of analytical data and advise farmers on adequate use of fertilizers
- Chemical and Organic fertilizer use
- Trash mulching
- Green manure crop

Irrigation

- Water needs of cane
- Method of application
- Sources of Water
- Water management
- Water conservation

Drainage

- Assessing levels i.e. use of piezometer
- Types of drains
- Levels of water to be maintained
- Conservation

Mechanisation

- Hand held equipments
- Animal drawn equipments
- Two wheel tractors and their use
- Four wheel tractors and their use
- Harvesters - Advantages and disadvantages.

Types of Harvesters

- Green cane harvester with full tracks
- Infield transporters
- Maintenance and servicing
- Off season use

Harvest and Transport

- Manual
- Mechanical
- Cane deterioration studies
- Extraneous matter: tops, trash, etc.

Extension Methods

- Advisory services
- Demonstration
- Audiovisual aids
- Training
- Preseason brixing and preparing harvest schedules.

Farmer Training

There are two categories

One day training

Four day training

Four day training

This consists of four-day residential training where in the farmers are given both theoretical lessons and on the job training. They are inducted into practicals of sugarcane farming, handling draft animals, two wheel tractors, four-wheel tractors use of various mechanical equipment etc.

They are given lessons in scientific aspects of cane plant, its care fertilisation, irrigation, crop protection and ratooning and overall crop production. Planning and finance management is also taught.

Crop Estimation

This is another very important aspect of the duties of the extension staff. First they will carry out the preseason brixing, i.e. testing all cane above eleven months for the sweetness of cane. Samples are taken and analyzed in the laboratory.

The harvest and transport officers are then informed of the fields that can be harvested. Thus the mill will receive sweeter cane for harvest.

Results and Discussion

The achievements of Parry through Research, Extension and Training has been large as can be seen from the factories data from the following locations:

Table 10.3:

Parameters	*Nelli-kuppam (1994)*	*Pugalur (1994)*	*Petta -vaitha lai (1998)*	*Pudukottai (2000)*
Cane	6,00,000	2,00,000	4,00,000	2,89,000
Sugar yield	56,000	18,700	21,300	24,800
No. of days crushed	150	80	157	128
Yield T/ac	36	33	38.4	34.0

2001-2002

Parameters	*Nelli-kuppam*	*Pugalur*	*Petta -vaitha lai*	*Pudukottai*
Cane	12,00,000	8,00,000	6,20,000	6,73,000
Sugar yield	113,200	75,400	56,600	63,900
No. of days crushed	280	200	266	254
Yield T/ac	43	38	39	35.4

There has been a substantial increase in crush without capital outlay. Yields have gone up substantially and so also recoveries of sugar. Better utilisation of assets and hence lower fixed costs.

The Agriclinics and Agribusiness Centres Scheme: An Innovative Extension Approach

Alexander George and S. Bhaskaran

ABSTRACT

The ambitious "Agriclinics" and Agribusiness Centres Scheme" was launched in April 2002. The programme is being spear headed by the National Institute of Agricultural Extension Management (MANAGE), in collaboration with the Department of Agriculture and Co-operation, Ministry of Agriculture, Government of India, National Bank for Agriculture and Rural Development (NABARD) and the Small Farmers' Agribusiness Consortium (SFAC). A network of about sixty-reputed training institutions all over the nation are involved in the actual work of imparting a series of two-month training programme for graduates in Agriculture and allied subjects so as to promote the establishment of Agriclinics and Agribusiness ventures. These include State Agricultural Universities, ICAR institutions, private firms and NGOs. Over 18,000 graduates in agriculture and allied subjects have registered for this training programme and over 2700 have been already trained. This concept paper deals with the newer and innovative approaches in the extension methodology, approach, procedure and strategies. It also analyses the unique features of the scheme, profile of trained personnel, profile of enterprises opted for by trained personnel and need for effective handholding mechanism if the scheme is to be effective. The suggested mechanisms for overcoming the constraints faced by the participants in the entrepreneurship development are also discussed.

Introduction

The ambitious "Agriclinics and Agribusiness Centres Scheme" was launched in April 2002. The programme is being spear headed by the National Institute of Agricultural Extension Management (MANAGE), in collaboration with the Department of Agriculture and Co-operation, Ministry of Agriculture, Government of India, National Bank for Agriculture and Rural Development (NABARD) and the Small Farmers' Agribusiness Consortium (SFAC).

This paper discusses: (1) The rationale behind the endeavour (2) Fears and apprehensions (3) Objectives and *Modus Operandi* of the scheme (4) Profile of trained personnel and the enterprises opted (5) Status of the scheme in Kerala (6) Need for effective handholding mechanism.

1. The Rationale Behind the Endeavour

Unemployed Agricultural Graduates: An Untapped Potential

In India, through 32 Agricultural Universities and 4 deemed universities around 11,900 Agricultural Graduates are passing out every year where as only 2000 are getting employment in government sector. Rest 9900 Agricultural Graduates go unemployed or search for jobs in the private sector or find avenues for self-employment. This has been happening for many years in the recent past. Out of the 1,10,000 existing public extension staff, only around 20 per cent are graduates. Is it possible to utilise the potential of this large reservoir of unemployed Agricultural Graduates in private extension?

Limitations of the Existing Public Extension System

The public extension worker—farmer ratio in India is 1:1000. This is far from sufficient. The changing roles and additional schemes entrusted on the public extension workers has resulted in an overload of administrative responsibilities preventing them from doing actual agricultural extension work. All said and done, the public extension officers often end up maintaining contact with a few progressive farmers who are easily accessible and mutually beneficial for the implementation of various developmental activities. Increasing budgetary constraints make it mandatory for alternative ways of financing agricultural extension services.

Changed Scenario

Gone are the days when a farmer depended on agriculture for food alone. Today farming is an enterprise and its goal is to produce marketable surplus to make profit and economic gain. The emergence of contract farming and the resultant changes in the cropping patterns adopted by farmers to earn more from smaller holdings has also set the stage for a more aggressive extension service. The implications of WTO on Indian agriculture make professional expertise indispensable for the Indian farmer to compete in the global economy. Technological advancements in the area of communication and information technology have made it possible today for extension activity to take place at a pace never possible a few decades ago.

All these factors in their synergistic consequence have already created a demand—a space that can effectively be filled by trained professional agripreneurs. The government is committed to develop private agricultural extension services. Having a parallel line of private professional agripreneurs can make space for the government functionaries to devote themselves more exclusively for priority issues of long-term significance and regulatory functions while leaving the regular extension work to the private agripreneurs.

2. Fears and Apprehensions

Will Farmers be Exploited?

One of the fears in promoting private agripreneurs is the fear that Indian farmers will be exploited. This fear is however not a cause of much concern as the Indian farmers have in the past been confronted with stray instances of being exploited by bogus and fraudulent private firms with respect to seed, pesticides and such other farming inputs. So there already exists in the farmers mind a sense of caution. Moreover with respect to services provided by private agripreneurs, it is expected that over the years a healthy competition between private agripreneurs and the public extension worker will come into existence as has happened in the medical, transport and telecom field. In the long run, the ultimate winner would be the farmer. He will decide the service provider based on the quality and reliability of the service he obtains to his level of satisfaction. He will also learn not to choose somebody who tries to exploit him. Licensing of agripreneurs and regulation by legislation can effectively prevent exploitation. The co-existence of public and private extension will help the ultimate consumer—the farmer through competitive, cost effective and efficient service.

Will Farmers Pay for Services?

An issue that has been raised by agripreneurs is: Does the Indian farmer have the ability to pay for extension services which hither to has been freely doled out by the public extension system? This fear is reasonable with respect to farmers in rain fed, unfavourable and resource—poor environments. Here public extension services will continue to have the major role to play.

Payment for services can however be looked at from a different perspective—the perspective of the dynamics that will come to play when services are paid for. When the farmer has to pay for services of an agripreneur, he will have a tendency to use the services/advice received more responsibly, because it is not free and because it has cost him his hard-earned money. Moreover the farmer has every right morally and legally to demand quality service. He commands respect and demands quality service because he pays. Willingness of the agripreneur to work beyond fixed hours and be available on call at any time will increase farmer satisfaction. From the entrepreneurs' perspective, the personal economic gain being inseparably linked to the professionalism that he has to offer will increase his involvement and challenge him to deliver better services.

Will the Farmer be Subject to Contradictory Messages?

Commodity oriented agribusiness and input dealers often try to push through their products. The competition among private agripreneurs may lead the farmers to confusion. Will the agripreneurs be as concerned for protection of the environment through integrated pest and disease management norms as the public extension system? This fear is reasonable and the only option seems to be to enhance the regulatory role of the public extension functionaries over the private agripreneurs.

3. Objectives and Modus Operandi of the Scheme

The main aim of the scheme is to *provide accountable extension services* to farmers through technically trained agriculture graduates at the village level. The scheme has accepted the challenge of *changing the attitude of agriculture graduates* from being job-consumers to job-producers. These efforts not only supplement the public extension services but also enhance the *quality of extension services* available to the farmers. Thereby the scheme also aims to *change the attitude of farmers* and motivate them to avail extension services through private for their prosperity.

There is a three-pronged focus in terms of change objectives: attitudes, accountability and quality. An attitude may be defined as a habit of thought. Among knowledge, skills and attitude, the three change targets of all extension work, attitude is the most difficult to change. Graduates have in the past always looked for a white collared secure job. They have never considered the possibility of being job producers rather than job consumers. Likewise, the act of asking for payment in terms of consultation fee or the work culture of being available at the farmers disposal even after official working hours require basic changes in ones own thinking and attitude.

In the changing global scenario wherein agriculture has become a competitive enterprise, the farmers need quality extension services. They need to be given the right information at the right time. The public extension services have not been able to rise to the level of expectations of the farmers in the changed competitive environment.

On the farmers part there must be attitude changes to be willing to change his own quality parameters to meet international standards and to pay the agripreneur for advice and services rendered.

Agriclinics are envisaged to provide expert services and advice to farmers on cropping practices, technology dissemination, crop protection from pests and diseases, market trends and prices of various crops in the markets and also clinical services for animal health.

Agribusiness centres are envisaged to provide input supply, farm equipment on hire and such other similar services.

The Agriclinis and Agribusiness Centres Scheme aims at training 25,000 agricultural graduates in five years at the rate of 5000 graduates per year. Over 18,000 graduates in agriculture and allied subjects have registered for this training programme and over 2700 have already been trained. A network of 60-reputed training institutions all over the nation are involved in the actual work of imparting a series of two-month training programme for graduates in Agriculture and allied subjects so as to promote the establishment of Agriclinics and Agribusiness ventures. These include State Agricultural Universities, ICAR institutions, Private firms and NGOs.

4. Profile of Trained Personnel and Enterprises Opted

As on March 2003, over 18,000 graduates in agriculture and allied subjects have registered for this training programme and over 2700 have already been trained. Interesting to note is that 31 of the 2712 candidates have trained are above 60 years of age and 117 of those trained are women.

Some of the enterprises that have had takers from amongst the trained personnel are: Agriclinics providing consultancy services and Agribusiness Centres with sales of inputs and services, Vermi-composting Units, Mushroom Cultivation Units, Seed Processing and sales, Landscaping and Nursery, Herbal Gardens, Organic Production/Food chains, Contract farming, Cyber Extension, Animal feed units, Farm Machinery on rent, Dairy, Poultry, Direct marketing, Agri-Eco Tourism, Tissue Culture Units, Soil Testing Units, Honey production units, Micro Irrigation Systems, Post Harvest Management and Food Processing Units, Dry flowers marketing

5. Status of the Agriclinics and Agribusiness Scheme in Kerala

In Kerala, the Central Training Institute, Mannuthy under the Kerala Agricultural University is the only center imparting the Agriclinics and Agribusiness Training. Three batches of the Agriclinics and Agribusiness Training have been completed as detailed below:

Batch No.	Dates of Participants	
1st	14th May 2002 to 15th July 2002	27

2nd	23rd October to 21st December 2002	30
3rd	29th July to 27th September 2003	30
	Total	87

The competency gained through the Agriclinics and Agribusiness Training and the Government of India certificate issued to trainees enabled some of the trained personnel got jobs in agribusiness and export firms either in India or in the Arabian Gulf countries. Some were motivated to pursue their post graduation so as to equip themselves sufficiently enough to provide specialised consultancy services in the competitive global scenario. Some of the others who ventured out on their own are:

Dr. P.P. Moosa [ID: KER 0016] who serves as a private professional consultant for cardamom and pepper farmers' in Vellamunda, Panavaram and Manathavady panchayats of Wyanad district, Kerala. Being a doctorate degree holder in Soil Science, and having worked at the Government Soil Testing Lab, Wyanad, Dr. P.P. Moosa is uniquely qualified to take soil samples and give recommendations. He also supplies quality vegetable and paddy seed. He purchases high quality tissue cultured Anthurium plants from the Government Tissue Culture Lab, Kazhakoottam, Thiruvananthapuram and sells them to the farmers at Wyanad. For these services farmers pay him the travel allowance plus 20% commission. Dr. P.P. Moosa sells tea dust and candles through a group of twenty-five women who are members of a Self-Help Group (SHG) at Malapuram. Through this exercise the SHG gains a profit of Rs. 2000/- per month and Dr. P.P. Moosa gets a profit of Rs. 1000/- per month. Dr. P.P. Moosa serves as a consultant for the Imam Gassaly Academy. Students of this academy grow plants for seed purpose and sell the seeds. The students take the profits. He is mobilising a group of farmers so that they could register under the Mridha Sanjeevini Scheme of Kottakal Ayurveda Shala where by these farmers will be in a position to supply medicinal plants easily available in Wyanad district. Land

has also been purchased for starting a Goat farm. A training centre for food processing, bee keeping, vermi-composting is also being planned.

Dr. E. Sivaraman [KER 0190] **has founded the Institute of Agricultural Sciences, Mannuthy registered under Company Act with Board of Directors and Advisory Committee. The Institute provides consultancy, training and communications. Two of the major clients of the institute are SILPI Agro Tech Pvt. Ltd. and Konenco Feeds It is also involved in:**

- **Monitoring the quality of raw material and product formulation of different kinds of feed**
- **Advising on up gradation of quality and purchase of micronutrients**
- **Organizing seminars on Livestock Management**
- Placement of Technical Personnel

Sri. K. Georgekutty Mathew [KER 0033] has mobilised a group of farmer entrepreneurs, got them registered as a Society under the name FLORALS. The society provides professional assistance in landscaping and lay out of house and office gardens in Kannur District of Kerala.

Smt. Raja Anupama Manavikraman [KER 0181] has started a Vanilla Nursery. Over 350 cuttings of Vanilla have been planted in a 40 cents plot and sales have commenced. She is also considering a tie up with some firms for marketing of Agricultural inputs.

Sri. Rajee C. George [KER 0126] has also started a Vanilla Nursery at Cherakaekode in Thrissur District.

Smt. Sreejaya K.C. [KER 0191] has established a Vermi composting Unit in Thrissur District.

Sri. Joseph John Therattil [KER 0134] has started the Swashraya Karsha Samithi AgriClinic, Machanthode,

Thachampara. Consultancy Services, Soil Testing, Sale of Pesticides, Weedicides and Hormones are undertaken.

Sri. Allan Thomas [KER 0205] and **Smt. Parvathy S.** [KER 0173] have started an NGO registered under Charitable Societies Act named AEGIS School of Agri-Consultancy to provide specialized assistance to agriculturalists

O.R.G.A.N.I.C.

The participants of the first batch of trainees have registered themselves into a Non-Governmental Organisation by the name O.R.G.A.N.I.C. (Organisation of Rural Growth Networking of Agricultural Information Centers)

A postgraduate student of the Kerala Agricultural University is presently doing an evaluation of the effectiveness of the Agriclinics and Agribusiness Training conducted at the Central Training Institute, Mannuthy.

6. Need for Effective Handholding Mechanism

It is not sufficient for the agripreneurs to be trained and left to work on their own. A comprehensive system of hand holding facility should be provided for those who have been trained at least for one year during pre-establishment and post-establishment period of the enterprise.

This should involve

1. Regular contact with the training institute by convening get-togethers to discuss progress and problems.
2. Helping agripreneurs to obtain the licences for input trade, procuring technology, software and machinery.
3. Assisting agripreneurs to start the enterprise with or without loan.
4. Facilitating loan sanction from the bank to the agripreneurs by establishing rapport between agripreneurs, the banker and NABARD.

5. Promoting networking among agripreneurs.
6. Organising interface meetings between agripreneurs, bankers, agribusiness companies, NGOs, State Government departments, etc.
7. Facilitating consultancy from technology generators: Agricultural Universities and ICAR institutions

For this purpose an amount of Rs.5000 per trained candidate must be provided in the scheme. Although initial provision was made for this, the provision has later been kept in abeyance, thus hampering all hand holding facility to the trained agripreneurs.

Conclusion

During the most challenging first year of the Agriclinics and Agribusiness Centres Scheme, MANAGE has been exceptionally aggressive in documenting and disseminating over 500 success stories. Most of these determined agripreneurs have established enterprises even without waiting for a loan. These 500 plus cases spread all over the nation stand to prove that the scheme has set the ball rolling. For farmers it was free extension services earlier but paid and accountable extension services are now making inroads. Agricultural graduates are now thinking in terms of being job producers rather than being job consumers. The challenge before us is to strengthen the hands of these path breakers and to pave the way for many more to follow suite.

References

Chandra Shekara, P. (Ed.) 2001.Private Extension In India. National Institute of Agricultural Extension Management, Hyderabad.

Chandra Shekara, P. and Hassan, S.M., 2003.Success Stories of Agripreneurs. National Institute of Agricultural Extension Management, Hyderabad.

Role of NABARD in Agriculture and Rural Development

Sahoo, B.B

ABSTRACT

Poverty is a curse on humanity. No country is free from this evil. However, the case of India is something different from other countries due to the magnitude and nature of poverty. Though poverty incidence among the rural population in the country has seen a decline over the years from 56.4 per cent in 1973-74 to 27.1 per cent in 1999-2000, in absolute term, the change is less impressive as 260 million persons in the country of which 193 million in rural areas are estimated to be below poverty line. Agriculture sector, which is the main stay of millions of rural people, has become expensive and modernized. Further, the fruits of the different government and non-government programmes do not reach the poor effectively/properly. In this context, this paper briefly analyses the very concept of rural development, the role played by NABARD in agriculture and rural development and emerging issues. Finally, the paper offers certain suggestions as to how to cope with the situation.

Introduction

"I would say that if the village perishes, India will perish too. It would be no more India. Her own mission in

the world will get lost. The revival of the village life is possible only when it is no more exploited".

(Mahatma Gandhi on 'Harijan', 29 August 1936)

Poverty is a curse on humanity. No country is free from this evil. However, the case of India is something different from other countries due to the magnitude and nature of poverty. Though poverty incidence in the country has seen a decline over the years from 56.4 per cent in 1973-74 to 27.1 per cent in 1999-2000, in absolute term, the change is less impressive as 260 million people in the country are estimated to be below poverty line. In this context, this paper briefly analyses the very concept of rural development, the role played by NABARD in agriculture and rural development and emerging issues. Finally, the paper offers certain suggestions as to how to cope with the situation.

I. Rural Development-The Concept

The very concept "Rural Development" is complex in nature. With the change of time, socio-economic development and world order, the meaning of rural development has been changing. Further, the definition of the term cannot be universalized as its meaning and significance differ from country to country and from region to region. However, for understanding the meaning of the term in Indian context, it may be defined as: -

"A process of change in attitudes, beliefs and customs of the rural people, which helps in modifying and dynamising a stagnant and traditional rural set up towards a self-sustained and well-equipped society (both social and economic) with a better foresight and standard of living of the masses".

And in the narrow sense it may be defined as *'the socio-economic development of rural areas and rural people'*.

An average Indian village

India is predominantly an agrarian country with a variety of socio-economic structure and cultural heritage.

There are about six lakh recognized villages in this country and each village has its own characteristics. The major characteristic features of an average Indian village are the following:

- Highly skewed distribution of land holding.
- Existence of landlords, petty shop keepers/ manufacturers/traders, contractors, cultivators, agricultural labourers, daily wage earners and artisans.
- Cultivation is the main occupation. Farmers have little knowledge on cultivation. The mode of cultivation is traditional and mono-crop is the age-old practice of the farmers, which is mainly due to the absence of irrigation facility.
- Many farmers and rural entrepreneurs, in spite of the improved institutional credit delivery system/mechanism, are not able to make proper use of the credit system due to their ignorance, inability to provide collateral security and less/ no interaction with the credit institutions.
- Inadequate knowledge about the available opportunities, communication system, medical facility, drinking water and electricity are some other impediments, which make the poor poorer.

Composition of a Village Community

The village community may be grouped into poor class and not so poor class of people taking land holding, job and asset position into account.

- Poor class—Semi-tiled/thatched house or no house, inadequate land holding or no land, inadequate/no education and in-adequate/no resources or job for daily living.
- Not So Poor Class—Owning pucca house, adequate cultivable land for living, i.e., owning

more than 5.0 acres of cultivable land, better education, permanent job and sufficient resources for daily living.

Characteristic Features of People Living in Villages

Poor class

- Either less land or no land, i.e., Small farmers, Marginal farmers, Landless Agricultural Labourers, Wage Earners, Artisans.
- Illiterate and ignorant mass, no/semi skill in different farm/non-farm activities
- Either less or no role in village Panchayats/ politics

Not so poor class

- Own land, wealth and resources
- Education, skill and jobs (both Govt. and Pvt. Sector)
- Important portfolios in Village Panchayats/ politics
- Easy access to banks, schools, block offices and other govt. offices
- Influence over the village community

Components of Rural Development

The basic components of rural development are infrastructure/linkage, credit flow, market and education.

- Infrastructure/Linkages—Inputs, capital, market, knowledge/skill and demand
- Credit Flow—Demand-pulled production, Transportation/communication and Income of the people
- Market—Demand-pulled production, Transportation/communication and Income of the people

- Education—To utilise local man-power, To utilise local resources, To upgrade skill and Knowledge on the facilities/help available by the different Govt./Non-govt. agencies

Development Agents of a Village

The development agents of a village are the following.

- Village Panchayat
- Rural Youth Force
- Local Resources
- Educational Institutions
- Non Government Organisations
- Rural Industries

What to do to Bring Prosperity in Villages

Development of Rural Areas

- Infrastructure/linkage development
- Easy and all-time access to nearby towns/ market centers
- Market development
- Credit Institutions
- Industries based on local resources/man power

Development of Rural People

- Education/skill upgradation
- Cooperation among the fellow partners
- Knowledge on Government programmes
- Easy and timely access to credit
- Better market/linkage for the products of farmers/entrepreneurs

How the Agents/Components will help in Bringing Development

The above agents/components will help in bringing a change in farm sector, non-farm sector and social sector.

- Agricultural Development—Mechanisation and Diversification
- Allied Sector Development—Dairy keeping, Fishery and related allied activities
- Non Farm Sector Development—Industries based on local resources/man power
- Improvement in Social Sphere—Improvement in Living condition/style, Education, General health, Awareness and Communication

Basic Parameters of Rural Development

- RD = f (I/L, E, Cf, M)
- E_a = f (VP, RYF, NGOs, EIs)
- C = f (VP, G, B, VCF)
- M = f (T/C, DPP, Yp)
- I/L_a = f (G, VP, NGOs)

Where, RD = Rural Development, f = function, E = Education, a = agent, Cf = Credit Flow, M = Market for farm/non-farm products, I/L = Infrastructure/Linkages, VP = Village Panchayat, RYF = Rural Youth Force, NGOs = Non Government Organisations, EIs = Educational Institutions, G = Government, B = Bank, VCF = Village Chit Fund (of SHGs), T/C = Transportation/Communication, DPP = Demand-polled Production, Yp = Income of the people.

II. National Bank for Agriculture and Rural Development

Genesis

The Agricultural Finance Committee (otherwise known as Gadgil Committee, 1945) had recommended for an autonomous Provincial Agricultural Credit Corporation, but it was negated by subsequent Committees like All India Rural Credit Survey Committee (AIRCSC-1954), All India Rural Credit Review Committee (AIRCRC-1969), Agricultural Refinance Corporation (ARC-1970) and Banking Commission (1972) on the ground that it would only add additional burden and red-tape without adding resources or efficiency to the farm sector. However, National Commission on Agriculture (1976) felt the necessity of a separate apex institution on agriculture and rural development outside the purview of Reserve Bank.

Government of India examined the recommendation of the Commission, i.e., in the year 1979, and advised Reserve Bank of India (RBI) to appoint a committee to look into the matter. A Committee in the name of CRAFICARD (Committee to Review the Arrangements of Institutional Credit for Agriculture and Rural Development) was then appointed in the same year (November 1979) to examine the whole gamut of agriculture, rural credit delivery system and rural development. The Committee in its interim report recommended for the establishment of an apex bank in the name of *National Bank for Agriculture and Rural Development* (NABARD). The proposal was approved by Government of India and Reserve Bank of India and NABARD Act, 1981 was passed by the Parliament. As a result, by amalgamating Agricultural Refinance and Development Corporation (ARDC), Agricultural Credit Department (ACD), and a part of Rural Planning and Credit Cell (RPCC), NABARD came into existence on 12 July 1982.

Main Objectives

The mandate of NABARD is rural prosperity. It has been created for providing undivided attention, forceful direction and pointed focus to the credit problems of the rural economy. As the apex institution, NABARD is concerned with policy, planning and operation relating to agriculture and rural development. The main objectives, which guide the functioning of the National Bank, are the following:

- To provide short-term, medium-term and long-term refinance to eligible institutions for supporting production and investment credit for developmental activities in rural areas.
- To take effective steps through institution building, restructuring of credit institutions, formulating and monitoring of suitable policy, etc., for improving the absorptive capacity of the credit delivery system.
- To coordinate the activities of different agencies engaged in agriculture and rural development.
- To keep a liaison among various institutions from bottom to top engaged in agriculture and rural development.
- To undertake monitoring and evaluation of projects refinanced by it.

Rural Development in NABARD's view

Development of Rural Area

- Providing required infrastructure
- Creating Market for the produced items
- Making the eligible Financial institutions related to the development of agriculture and rural development strong, efficient, viable and people-friendly

Development of Rural People

- Providing easy credit facility
- Skill upgradation/improvement through training
- Creating Awareness among the village people
- Encouraging people to take the available opportunities for their development

Functions

The major functions of NABARD can be split into Credit function and Non Credit function.

Credit Functions

Functions relating to Financial Services

- Refinancing Rural Financial Institutions for Production and Investment purposes in rural areas.
- Loans to State Governments for developing rural infrastructure and strengthening cooperatives.
- Support for credit innovations of Non Governmental Organisations and other non-formal agencies.
- Monitoring and Evaluation of financed projects.

Functions relating to Credit Planning and Monitoring

- Preparation of district-wise Potential Linked Credit Plans annually, which indicate exploitable potential under agriculture and allied activities available for development through bank credit.
- Monitoring the flow of ground level credit.
- Issuing policy and operational guidelines to Rural Financial Institutions.

Investment-wise Refinance Disbursement

During the last twenty-one years, i.e., 1982 to 2002, the refinance disbursed had increased manifold. The cumulative total refinance disbursement by NABARD by the end of the year 2002-03 was Rs. 1,67,073 crore. Investment-wise annual increase in the refinance disbursement was the maximum of 38.91 per cent for medium term (including conversion) followed by short term investment credit, i.e., 22.46 per cent and the annual increase in the refinance disbursement during the years 1983 and 2003 was 38.35 per cent. (See Annexure I)

Purpose Wise Refinance Disbursed by NABARD

The major purposes covered under Medium Term and Long Term investments are Farm Mechanisation, Minor Irrigation, Land Development and Non Farm Sector. The annual increase in refinance disbursement during the years 1983 and 2003 was the maximum for animal husbandry/ dairy development (600.38 per cent) followed by the programmes under IRDP/SGSY (165.49 per cent) and farm mechanisation (99.51 per cent). (See Annexure II-A).

When the investment credit of NABARD is split up into farm sector and non-farm sector, it is found that investments in both farm and non-farm sectors had increased over years, but the increase in investment in non-farm sector (214.97 per cent) was more than the investment in farm sector (47.77 per cent). (See Annexure II-B).

Agency Wise Disbursement of Refinance

The annual increase in the refinance intake of the Regional Rural Banks during the years 1983 and 2003 was 429.66 per cent followed by the State Cooperative Banks, i.e., 282.07 per cent and the Commercial Banks, i.e., 55.39 per cent. (See Annexure III).

Region Wise Refinance Disbursement

When the refinance disbursement of NABARD is studied over regions, it is found that during the years 1983 and 2003, the annual disbursement of refinance was the maximum for North Eastern region followed by Southern region and the minimum for Northern region. (See Annexure IV).

Non-credit Function of NABARD

To strengthen the credit delivery system and to make the primary institutions operationally efficient, credit is not the only requirement. These primary institutions should be strong, efficient and financially viable for undertaking effective credit delivery. Further, the National Bank to perform the functions of an apex institution must have some regulatory control over its client banks to get its policies implemented. Understanding its importance NABARD performs a number of non- credit functions. The non-credit functions of the National Bank can be grouped under four heads, i.e., Developmental[1], Promotional[2], Statutory & Regulatory[3] and Other Functions[4].

Overall Performance of NABARD

The overall performance of NABARD in terms of funds, business, income and profit during the year 1983 and 2003 is found to be very satisfactory. The annual increase in business, income and profit of the organisation was 38.28 per cent, 72.61 per cent and 53.34 per cent respectively. (See Annexure V).

Critical Assessment of the functions/activities

NABARD, as an independent body, came into picture in the year 1982 and since its inception, it has significantly been contributing for agriculture and rural development in the country by providing an ideal infrastructure base. It has successfully completed 21 years of service to the nation. The traditional function of the Bank was to supplement the resources of the Cooperative Banks, Regional Rural Banks and Commercial Banks by providing refinance support against their lendings at the grassroots level for expanding and diversifying the ground level credit for agriculture and rural development as also for directing credit to the thrust areas prioritized in the Five Year Plans/Annual Union Budgets of Government of India from time to time. During these 21 years, the Bank has taken rapid strides in all the traditional functional areas and at the same time it has also taken up new challenges.

Recent programmes like Self Help Groups[5] (SHGs), Rural Infrastructure Development Fund[6] (RIDF) and Kisan

Credit Card (KCC) have become mass movements, touching the lives of millions of people. Recognizing inadequate storage capacity as one of the major infrastructure bottlenecks of agricultural reforms, NABARD in collaboration with Government of India has been encouraging investments in construction, modernisation and expansion of cold storage and rural godowns. NABARD has also been encouraging private investment in other infrastructure like setting-up of market yards. The gap between lab and land in transfer of technology has been found to be a hurdle in the adoption of better farming practices. To bridge this gap, the Bank has designed a scheme in consultation with Union Ministry of Agriculture for setting-up of agri-clinics and agri-business centers by agricultural graduates as a step towards privatisation of agricultural extension system for enhancing technology dissemination.

A beginning has also been made in direct lending through co-financing with financial institutions. In July 2002, NABARD Consultancy Services (Nabcons) Cell was also set up in the Bank to provide services to various client institutions and individual investors. Besides, as a step towards encouraging scholars for knowledge dissemination on agriculture and rural development, strengthening the rural financial institutions, financing promotional and developmental activities for rural non-farm sector, spreading the message of participatory watershed development throughout the country and promoting and strengthening Self Help Groups, NABARD extends grant support under Research & Development Fund, Cooperative Development Fund, Credit and Financial Services Fund and Rural Promotion Corpus Fund, Watershed Development Fund and microFinance Development Fund respectively.

The traditional functions:

- Supplements the resources of the cooperative banks, RRBs and CBs by providing refinance support against their lendings for expanding and diversifying the ground level credit for agriculture and rural development,

- Directs credit to the thrust areas prioritized in the Five Year Plans/Annual Union Budgets of Government of India from time to time.
- Encourages scholars for knowledge dissemination on agriculture and rural development

New Functions as a change agent:

- Builds infrastructure in irrigation, roads and bridges, soil conservation and watershed management under Rural Infrastructure Development Fund.
- Facilitates relationship banking through Self Help Groups by the rural poor.
- Strengthens credit delivery system at the grassroot level by Kishan Credit Card Scheme
- Encourages investments in infrastructures like market yards, cold storages and rural godowns.
- Encourages privatisation of in the field of agriculture sector by launching Agri-clinics and Agri-business (in extension service), co-financing with financial institutions for hi-tech, export-oriented agricultural projects (direct financing) and setting up of Agri-Export Zones.
- Encourages farmers to adopt 'Contract Farming' for economies of scale and better returns on their produce/commodity.
- Provides refinance support for rural housing to houseless poor.
- Encourages for the development of non-farm sector based on local man-power and local

resources through Rural Entrepreneurship Development Programme.

- Encourages participatory watershed development through the Watershed Development Programme.
- Provides Consultancy services to various institutions in the field of agriculture and rural development.

III. Emerging Issues in the field of Agriculture and Rural Development in the Changing Scenario

Agriculture sector has the potential for reducing poverty and hunger, so this sector cannot be neglected. It is a concern that cultivable land, fertility of land and land holding size are declining. Further, population size, domestic demand pattern and external market are fast changing. Under such a situation, agricultural output needs to be increased at a manifold rate keeping quality and cost of production fairly competitive to meet the demands of both internal and external markets. When compared to international standards, our productivity is low and cost per unit in general is higher despite lower wages. Unless Indian agriculture is competitive in quantity, quality and cost, imports would increase and export opportunities would fade out. The major problems are the following:

- Net area devoted to agriculture is shrinking.
- Landholding size is becoming smaller and smaller.
- Wasteland is on the increase.
- Pollution of water and degradation of land and soil is increasing.
- Urban-rural divide and regional disparities are on the increase.
- Contribution of agriculture to Gross Domestic Product is declining.

- Per capita availability of food has not increased significantly in spite of higher production.
- Demand for water by various user interests is on the increase, which calls for precision use of water.
- Cheap machine-made products outplace the rural artisans and they were forced to become casual labourers.

IV. Suggestions

- The solution to many of the problems raised above lies with the very concept of Self Help Group.
- If the poor are provided with opportunity, guidance and credit plus service, they will definitely come out of poverty, inequality and ignorance.
- The unity and fund raising ability among the participant members in the group will help them in modernizing their age-old agricultural practice, changing their cropping pattern in favour of remunerative cash crops and making efficient use of their fragmented and small-sized production base through group-farming and contract farming.
- It will also encourage the farmers to make use of the available wasteland and fallow land in their locality.
- Proper guidance to the farmers/members of the group will encourage making efficient use of the scarce water available through the use of water saving devices like sprinkler.
- Credit opportunity will help rural women and men to take up the new role of entrepreneurs

by making use of the available local resources and demand, which will empower them economically and socially.

- Group approach would promote collective action, responsibility and knowledge sharing in different fields including agriculture.
- It would also lower the cost of cultivation.
- However, for effective translation of the approach, care must be taken by different agencies for the qualitative improvement of the groups.
- Both backward and forward linkages in the form of basic inputs, market and price must be created to attract the farmers.
- Private bodies may be attracted into Agri-business, Corporate farming and Contract farming in order to increase investment in agriculture sector in a big way. Knowledge dissemination, skill upgradation and grounding of laboratory experiments on farm sector need expansion.
- Non-farm sector along with farm sector must be given priority for generating meaningful employment and income for the rural women, men and the deprived class of people, which will supplement the growth of agriculture sector in the country.

Conclusion

Both top-bottom and bottom-top approaches have been experimented to combat poverty, unemployment, illiteracy and malnutrition. However, the success achieved falls short of the need. It has also been felt that rural development is not in "providing" but in "promoting" the rural sector. Therefore, effective and self-sustaining measures are needed for the rural uplift. Credit is not the only requirement for achieving the goal of rural development. Among others, political will, public participation and commitment are the key factors.

I, personally, feel that agriculture and rural development would not be a distant goal, if we collectively work with the present Self Help Group Movement. This concept can be replicated in all the sectors/fields of rural development.

REFERENCES

NABARD, Annual Report, Various Issues

Reserve Bank of India, "Reserve Bank of India: Functions and Working", 1984.

Reserve Bank of India, "Report of the Committee to Review Arrangements for Institutional Credit for Agriculture and Rural Development (CRAFICARD), 1981".

Government of India, "Report of the National Commission on Agriculture, Government of India", 1976, Part XII.

Tendulkar, Suresh D. Rural Institutional Credit and Rural Development, A Review Article, Indian Economic Review, Vol.XVIII, No.1, Jan-June, 1983.

NABARD, "Initiatives and Performance, 1982-2003".

Gupta, S.C. Development Banking for Rural Development, Deep & Deep Publications, New Delhi, 1987.

NABARD, "Rural Credit & NABARD 2003".

Reserve Bank of India, *"All India Rural Credit Review Committee Report"*, Bombay, 1969, p. 411.

NABARD, *"Report of the Expert Committee on Rural Credit"*, 2001.

ANNEXURE I

Details of Investment-wise Refinance Disbursement Position (Cumulative) during 1983 and 2003

(Rs. in crore)

Particulars	*1983*	*2003*	*%Increase*		*Cumulative Total by 2003*
			Total	*Annual*	
Short Term Investment Credit	1231	7038	471.73	22.46	83535
Medium Term (including Conversion)	811	7438	817.14	38.91	65440
Long Term Loan to State Government	13	28	115.38	5.49	954
Total MT/LT Credit	824	7466	806.07	38.38	66394
Rural Infrastructure Development Fund	-	4103	-	-	17144
Total	2055	18607	805.45	38.35	167073

ANNEXURE II-A

Details of Purpose-wise Refinance Disbursement (Cumulative) during 1983 and 2003

(Rs. in crore)

Particulars	*1983*	*2003*	*% Increase*	
			Total	*Annual*
Minor Irrigation	1921	12543	552.94	26.33
Farm Mechanisation	664	14539	2089.61	99.51
Plantation & Horticulture	146	2817	1829.45	87.12
Animal Husbandry/ Dairy Development	62	7879	12608.06	600.38
Non Farm Sector/ Rural Housing		9146		-
IRDP/SGSY	300	10726	3475.33	165.49
Self Help Group		1413		-
Others**	455	5921	1201.32	57.21
Total	3548	64984	1731.57	82.46

** *Includes Land Development, Poultry/Sheep/Piggery, Fisheries, Storage & Market Yards, Forestry, Biogas and others.*

ANNEXURE II-B

Details of Purpose-wise Refinance Disbursement (Cumulative) during 1983 and 2003

Particulars	*Amount (Rs. crore)*		*% Increase*	
Farm Sector	2812(79.26)	31022 (47.74)	1003.20	47.77
Non Farm Sector	736(20.74)	33962 (52.26)	4514.40	214.97
Total	3548(100.0)	64984 (100.0)	1731.57	82.46

Figures in brackets refer to percent to total for the respective years

Investments in Minor irrigation, Land development, Farm mechanisation and Plantation and Horticulture are included under farm sector investments and the investments in other sectors are included under non-farm sector category.

ANNEXURE III

Details of Agency-wise Refinance Disbursement (Cumulative) during 1983 and 2003

(Amount in crore)

Particulars	*Amount (Rs. crore)*		*% Increase*	
SLDBs/SCARDBs	1656.7	26741.57	1514.20	72.10
Commercial Banks	1663.9	21018.14	1163.18	55.39
State Cooperative Banks	118	7107.61	5923.40	282.07
RRBs	109.65	10003.26	9022.90	429.66
ADFCs/PUCBs	-	113.13	-	-
Total	3548.2	64983.71	1731.45	82.45

ANNEXURE IV

Details of Region-wise Refinance Disbursement (Cumulative) during 1983 and 2003

(Amount in crore)

Particulars	*Amount (Rs. crore)*		*% Increase*	
Northern	838.63	13463.66	1505.44	71.69
North Eastern	32.89	1295.48	3838.83	182.80
Eastern	428.14	7262.24	1596.23	76.01
Central	803.2	13786.32	1616.42	76.97
Western	511.23	10089.65	1873.60	89.22
Southern	925.69	19086.31	1961.85	93.42
UT's	8.43	-	-	-
Total	3548.2	64983.66	1731.45	82.45

ANNEXURE V

Details of NABARD's Performance (Cumulative) during 1983 and 2003

(Amount in crore)

Particulars	*Amount (Rs. crore)*		*% Increase*	
Working Funds	4519	50885	1026.02	48.86
Owned Funds	1922	20738	978.98	46.62
Total Business	2055	18573	803.80	38.28
Income	249	4046	1524.90	72.61
Profit	94	1147*	1120.21	53.34

Profit for 2002-03 excludes Income Tax of Rs. 377 crore.

13

Extension Approaches of State Bank of Travancore for Empowerment of Rural Sector

Subramoniam, K.

ABSTRACT

State Bank of Travancore, the premier bank of Kerala is closely associated with rural development for the last four decades. Apart from participating in the implementation of poverty alleviation programmes of the government, the bank is helping the rural poor through its loan products under agriculture, small business segment and small-scale industries sectors. Some of the products are Kissan Credit Card, Kissan Gold Card, and loans for land development, minor irrigation, farm mechanisation, plantation and horticulture, animal husbandry, fisheries, beekeeping, sericulture, artisans, Lakhu Udhyami Credit Card (SBTLUCC) etc. This paper entails the techniques followed by the bank in propagating the rural lending schemes and other activities. The techniques followed by the bank have been found to be very effective tools.

Introduction

The problem of poverty is the most formidable one facing independent India. World Development Report (WDR) 2000-2001 reveals an interesting fact. Eventhough India's GNP is 11th largest, as much as 35 per cent of India's population is below the national poverty line and

44.2 per cent is below the international poverty line (earning $ 1 per day).

Several factors account for the poverty of Indian population, the important among these are as follows:

First, most of the people in the rural areas do not own any land or any other productive asset. There is positive relationship between poverty and absence of ownership of income-earning assets. Secondly, low productivity of land and labour is also responsible for poverty. Thirdly, high rate of population growth and lack of corresponding new job opportunities are also contributing factors for poverty. In the absence of alternative occupations in rural areas the number of unemployed persons is swelling rapidly. The more distressing feature of the agricultural labour is that they do not get employment all through the year. The wages too are miserably low, as a result, they have to lead a degraded life. Finally, poverty itself begets poverty. The three dimensions of poverty are income poverty, health poverty and education poverty, one leading to other making a vicious circle.

Data relating to national income growth depict that in spite of the higher priority accorded to agricultural development during the plan period, the percentage share of agricultural sector in the national income is continuously decreasing. Percentage share of agriculture and allied activities in national income has dropped from 55 per cent in the period 1951-56 to 33 per cent in the period 1985-92 and further to 27% in 2001-2002.

An important feature of Indian economy is that agriculture and industry are interdependent. Agro-based industries get their raw materials from agriculture and agricultural sector is the largest buyer of industrial products like fertilizers, pesticides, mechanised implements, irrigation equipment, etc. Therefore, if agriculture fails, it will jeopardize the growth of other sectors of the Indian economy. Future development of the Indian economy rests upon the overall growth of different sectors.

The Role of Banks in the Development of Rural Sector

The concept of social banking took birth in the year 1967 and the year 1969 witnessed the major event of nationalisation of 14 commercial banks. The decade starting 1970 witnessed phenomenal changes in the banking system in India and it has evolved into a powerful instrument for regulating and directing the economy. Government of India set goals for banks for the achievement of higher agricultural growth by (a) making small farms more productive, and (b) raising the levels of incomes and employment in rural areas. In order to fulfill these purposes, banks were asked to ensure that credit would not be a constraint for development and to soften credit terms for priority sectors, particularly for the weaker sections of the community. This process might involve sacrifice to a certain extent of commercial principles of profitability and efficiency and bringing the socio-economic considerations to the fore. As a result, there was an unprecedented extension of banking facilities to the rural and semi-urban areas. The flow of institutional credit to the social classes who were earlier bypassed—small and marginal farmers, artisans, landless labourers, and unemployed youth—had become a significant feature of the rural credit system.

Economic Components of Rural Sector in India

Rural sector of India essentially consists of the following economic components:

(i) Agriculture and allied activities.

(ii) Small scale industries, ancillary units and cottage and village industries.

(iii) Retail trade and small business.

(iv) Small road and water transport operators.

(v) Professional and self-employed persons.

Rural Development in India—Efforts over Years

An analysis of the distribution of assets in rural areas indicates that the poorest 30 per cent of the population have practically no asset base. The basic feature of agrarian structure is that 73 per cent of the operational holdings in the country are cultivated by small and marginal farmers having only about 23 per cent of the total cultivable area. All technological changes, improved irrigational practices and additional inputs mainly reached the bigger operational holdings. Therefore, the first challenge in 1980's was to enable the small and marginal farmers and agricultural labourers to acquire or increase their productive assets so that their interest in farming and allied activities can be sustained.

Following the recommendations of All India Rural Credit Survey Review Committee, Small Farmers Development Agencies, Desert Development Agencies, Drought Prone Area Programme, etc. came into being. These agencies were linked to institutional credit to help in generating continuing income to the small farmers and the other vulnerable groups.

The Institutional Credit Policy of the Sixth Plan revolved around the following major objectives:

(a) to secure an increase in the total volume of institutional credit for agriculture and rural development;

(b) to direct a larger share of the credit to the weaker sections;

(c) to reduce the regional imbalances in the availability of credit;

(d) to bring about greater co-ordination between different credit institutions under the multiagency systems and

(e) to improve the recovery of institutional loans to ensure continuous recycling of credit.

For accelerating rural development, the strategy and the methodology adopted were:

(a) increasing production and productivity in agriculture and allied activities.

(b) Resource and income development of vulnerable sections of the rural population through development of the primary, secondary and tertiary sectors;

(c) Skill formation and skill upgrading programmes to promote self and wage employment amongst the rural poor;

(d) Facilitating adequate availability of credit to support programmes taken up for the rural poor;

(e) Promoting marketing support to ensure reasonable prices for products and to insulate the rural poor from exploitation in marketing of their products; and

(f) Provision of additional employment opportunities to the rural poor for gainful employment through various employment generation programmes.

In 1975, the Prime Minister Mrs. Indira Ganghi outlined a new-economic 20-point programme which precisely aimed at the development of weaker sections of the community. This programme did not receive the emphasis it deserved in the subsequent years due to political reasons.

IV.1 Integrated Rural Development Programme

It was only during the 1980s that a clear perception emerged and specific strategies have been outlined for enabling the target groups to cross the poverty line. It was recognised that households below the poverty line will have to be assisted through an appropriate package of technologies, services and asset transfer programmes as an essential step for raising the productive potential of the rural economy. The Integrated Rural Development Programme (IRDP) was introduced at this point of time as a powerful strategy for combating rural poverty. There was a shift in approach from individual to family with the formulation of this programme.

For almost two decades, it remained as the largest poverty eradication programme of the world. But it also could not achieve the main objective of bringing rural mass above poverty line. Hence a new programme, Swarna Jayanthi Gram Swarozgar Yojana (SGSY) was launched effective from April 1999 merging six programmes viz. Integrated Rural Development Programme (IRDP), Training Rural Youth for Self-Employment (TRYSEM), Development of Women and Children in Rural Areas (DWCRA) Supply of Improved Tools to Rural Artisans (SITRA), Ganga Kalyan Yojana (GKY) and Million Wells Scheme (MWS). The main objective of SGSY is to bring the assisted families (Swarozgaris) above poverty line within three years, by providing them with income generating assets through a mix of bank credit and government susbidy. The family-oriented approach for development gave way to cluster approach which is more rational in view of balanced growth of rural sector.

IV.2. Employment Generation Programmes

As a part of the efforts to eradicate poverty and generate employment opportunities, many programmes were evolved. They are briefly mentioned below:

(a) National Rural Employment Programme (NREP) The objective of NREP was to provide additional employment opportunities to the rural poor for gainful employment during the lean agricultural season.

(b) Self-employment for Educated Unemployed Youth (SEEUY) The aim of this scheme is to

extend loans to educated unemployed youths for their self employment. The borrower under the scheme should have passed atleast 10th class or he may be I.T.I. trained, in the age group of 18-35 years, living in an area with population less than 10 lakhs. His family income should be less than Rs. 10,000 p.a. Assistance under the scheme was provided for three types of ventures, namely, industrial ventures (up to Rs. 35,000/-), service ventures (up to Rs. 25,000/-) and business ventures (up to Rs. 15,000/-).

(c) Prime Minister's Rozgar Yojana (PMRY). Withdrawing all other self-employment programmes, Govt. of India introduced Prime Minister's Rozgar Yojana (PMRY) from 2nd October 1993, widening the scope of the scheme. The investment ceiling was enhanced to Rs. 1,00,000/- for all types of ventures keeping in view of cost escalation in project implementation. Now this is the biggest self-employment programme of the country.

IV.3. Self-Help Groups (SHGs)

This is the latest concept in empowering rural sector. SHG represents "a registered or unregistered group of people having homogeneous economic background, coming together voluntarily to save small amounts regularly and mutually agreed to contribute to a common fund and to meet their needs on mutual help basis." The group members use collective wisdom and peer pressure to ensure proper enduse of credit and timely repayment thereof. Less transaction cost and prompt repayment are the two important benefits to financing banks.

Role of State Bank of Travancore in Empowering Rural Sector

State Bank of Travancore, the premier bank of Kerala is closely associated with rural development for the last three and a half decades. The Bank is committed in the implementation of poverty eradication programmes and self-employment programmes of the government mentioned elsewhere.

Bank is providing help to the rural sector by extending different loan products to Agricultural Sector), Small Business Segment, Small Scale Industries Sector, Rural Artisans, Small Road and Water Transport Operators, and Professional & Self-employed Professionals.

V.1. Agricultural Sector

Under Agricultural Sector, the Bank provides

- Kissan Credit Card (short term crop loan)
- Kissan Gold Card (long term development loans)
- Land Development (Levelling, bunding, bench terracing, etc.)
- Minor Irrigation (wells, pumpsets, sprinklers, drip irrigation systems, etc.)
- Farm Mechanisation (tractors, tillers, threshers, rubber rollers, etc.)
- Plantation and Horticulture
  - *(a)* Rubber, coconut, cardamom, etc. (long duration crops)
  - *(b)* Medicinal plants
  - *(c)* Vanilla cultivation
  - *(d)* Fruits and vegetables
  - *(e)* Ornamental plants like orchids, anthuriam, etc.
- Animal Husbandry (milch animals, goats, poultry, piggery and duck rearing)
- Fisheries (mechanised boats, catamaram, inland pisciculture, etc.)

- Beekeeping
- Sericulture

V.2. Small Industries and Business Sector

Under this sector, Bank has the following products:

- Project finance for small scale industries
- Working capital finance for SSIs and retail traders
- Laghu Udhyami Credit Card (SBTLUCC)
- Finance for mini restaurants
- Composite loans to rural artisans and village and cottage industries
- Transport vehicles to small road and water transport operators (up to 10 vehicles)
- Tourism (eg. Houseboat loans)

V.3. Professional and Self-Employed Persons

- Loans for setting up of clinics to doctors
- Vehicles for doctors
- Office equipment, computers, etc. for computer professionals, lawyers, chartered accountants, etc.

Extension Techniques of State Bank of Travancore

Bank's lending schemes, particularly for agriculture, animal husbandry, fisheries, land development, artisans, village and cottage industries, small road transport operators and small business have special significance in so far as they provide the economic life blood to those who were deprived of Bank's lending facilities for long. Eventhough easy availability of money is one of the important factors contributing to agricultural and rural development, still more important is its effective use together with the latest relevant technology. This is the responsibility the Bank has undertaken.

In fact, neither money nor technology is the supreme. Supremacy lies in the persons who make use of them. It is essential for the Bank that efforts be made to promote rural lending on scientific lines to ensure its healthy growth and better impact.

From the model of extent of credit use (Fig. 1), it is evident that different factors such as social and personal, psychological, economic and environmental and communication methods influence the extent of credit use. Towards this, effective communication of the latest relevant technology and loan products and development of extension services can play highly positive role. Keeping this in view, State Bank of Travancore has adopted the following extension techniques:

1. Arranging farmers' meet at branch level.
2. Seminars of progressive farmers at panchayat, block, taluk or district level with the co-operation of Lead Bank, State Agricultural Department, FACT, etc.
3. Participation in exhibitions in agricultural melas organised by State Agricultural Department and Agencies like UPASI, etc.
4. Organising Vikas Voluntary Vahini (VVV) or Farmer's Club. They will serve as brand ambassadors of the Bank.
5. Formation, nurturing, training and financing of Self Help Groups. "What is not possible individually can be possible collectively." SHGs are the best vehicles for rural empowerment.
6. Participation in the programmes implemented by Vegetable and Fruit Promotion Council, Kerala.
7. Encouraging farmers to take up projects for which subsidy is available from National Horticultural Board.

8. Display of posters and banners showing loan products in the branch premises and other vantage points.
9. Field level contact of farmers and entrepreneurs by Branch Managers and Field Officers of the Bank.
10. Conducting special programmes with the co-operation of development agencies like Block Development Office, State Agricultural Department, District Industries Centre, KVIC, NABARD, etc.
11. Implementation of poverty eradication and employment generation programmes of Government like SGSY, PMRY, REGP, etc.
12. Financing of Agricultural Clinics set up by agricultural graduates.
13. Radio/TV talk by Bank officials explaining the various rural lending schemes of the Bank.
14. Advertisement of loan products through leading dailies and periodicals.
15. Arranging Entrepreneurial Development Programme (EDP) for rural artisans and beneficiaries assisted under Rural Employment Generation Programme of KVIC.

Conclusion

Empowerment of rural sector is the social responsibility of State Bank of Travancore. The extension techniques mentioned elsewhere are found to be very effective tools for the Bank to contact clientele, to arouse their interest and aspiration, to providde them the know-how of the useful innovations and motivate them to adopt innovative practices which will benefit them. The Bank is always in the forefront drawing inspiration from the words of our President H.E. Dr. A.P.J. Abdul Kalam—"Dream, dream and dream, convert dreams into thoughts and thoughts into action—aiming low is a crime."

SOCIAL and PERSONAL

Age, education, knowledge of latest relevant technology. Pressing family needs, knowledge of credit sources, social participation.

Economic and Environmental

Size of holding marketing facilities, Availability of inputs, storage facilities, Market prices, etc.

Psychological

Attitude towards credit agency and purpose of Credit, confidence in science and technology, Risk bearing capacity, change proneness, Achievement motivation, market orientation.

COMMUNICATION METHODS

Personal contact and marketing of loan products, Advertisements through mass media, correspondence, Telephone calls, empathic listening etc.

Fig 1. Model of Extent of Credit Use

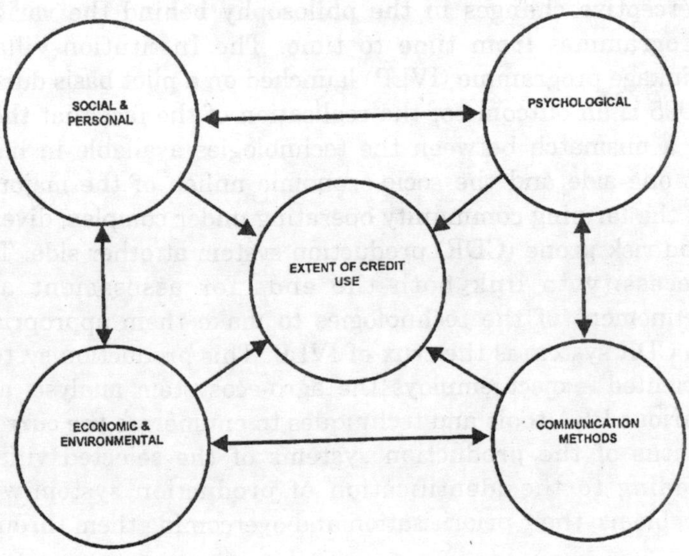

Institution-Village Linkage Programme—An Alternate Strategy for Technology Assessment and Transfer

Ramanathan, S. and Ananatharaman, M.

ABSTRACT

The understanding of the complex process of technology transfer from the lessons gained by way of implementing various front-line extension efforts have brought in conceptual changes in the TOT process. Starting from the National Demonstration programme to the latest Institution-Village Linkage programme, there were perceptive changes in the philosophy behind the various programmes from time to time. The Institution-Village Linkage programme (IVLP) launched on a pilot basis during 1995 is an outcome of the realisation of the fact that there is a mismatch between the technologies available in hand at one side and the socio-economic milieu of the majority of the farming community operating under complex, diverse and risk prone (CDR) production system at other side. The necessity to link both the ends for assessment and refinement of the technologies to make them appropriate to CDR system is the crux of IVLP. This production system oriented project employs the agro-ecosystem analysis and various PRA tools and techniques to enumerate the current status of the production systems of the selected village leading to the identification of production system-wise problems, their prioritisation and overcoming them through possible technological interventions in the form of on-farm trials or verification trials. This programme implemented by a multidisciplinary core team of scientists envisages active participation of farmers right from the beginning. During the course of implementation of IVLP, it was possible to categorize the technological interventions based on the adoption pattern. They are interventions requiring no precondition for adoption, adoption of interventions possible only with preconditions and the third category that is effective only at community level. IVLP as a whole was useful in identifying location-specific technologies and facilitated their faster spread. This approach need to be institutionalized and taken up by the agencies like research stations, KVKs, voluntary agencies etc. as a vehicle for rural development of our country with agriculture as the major occupation.

Introduction

Since Independence, various rural development programmes with main focus on agriculture had been formulated and implemented in successive five year plans. Though in the 1950s, agricultural development formed a part of the Community Development Programme encompassing various socio-economic development programmes in the rural areas, considering the food shortage coupled with population explosion on one hand and the advent of technological breakthrough taking place in agriculture on the other resulted in the creation of intensive and specific programmes of agricultural development. Ever since that period, a variety of programmes have been given shape and implemented. The understanding of the impact and experience gained under one programme led to the other which is different in concept and execution. It is this process which has now led to the Institution-Village Linkage Programme shortly referred as IVLP, based on the fact that the technologies are not scale neutral and do require detailed assessment and refinement needed if any to identify their appropriateness to small farmer production systems.

Agricultural Development Programmes

Starting with the Intensive Agricultural District Programme (IADP) launched during 1960, a series of such programmes followed one by one. The important amongst them are Intensive Agricultural Area Programme (IAAP) in 1964, High Yielding Varieties Programme (HYVP) in 1965, Multiple Cropping Programme in 1967-68, Special Programmes for Commercial crops etc. All these programmes by and large focused on the combination of all the technological improvements and concentration of manpower in selected areas which had optimum conditions for stepping up production.

Consequent of these intensive efforts, India witnessed the "Green Revolution" with phenomenal increase in the production of food grains especially rice and wheat. Notwithstanding this, there was also a severe criticism that green revolution made rich becoming richer and poor becoming poor. As all these programmes were concentrated in resource rich areas and farmers, a large majority of small and marginal farmers and landless labourers remain unattended. Moreover, the multi-various problems of problem areas were also untouched. This realisation paved the way for conceptual changes in the agricultural development programmes undertaken afterwards. Thus came the Drought Prone Area Programme, Dry land Agricultural Development, Small Farmers Development Agency and Marginal Farmers and Agricultural Labourers Development Agency, Command Area Development Programme, Hill Area Development Programme, Tribal Area Development Programme etc.

Front-line Extension Efforts

In line with the agricultural development programmes mainly carried out by the development departments on a large scale, Agricultural Research Institutions and State Agricultural Universities are also engaged in such activities in a limited scale, what is termed as front-line extension programmes. Starting from the National Demonstration Programme to the latest Institution-Village Linkage Programme, a series of programmes have been in operation from time to time.

National Demonstration Programme

The National Demonstration Programme was initiated in 1965-66 to demonstrate convincingly to farmers and extension workers the genetic potential of major crops per unit area of land, per unit time and encourage them to adopt/popularize these technologies for accelerating production. This programme followed a targeted approach and successfully shown the production potentialities of the new varieties and cultivation practices of cereals, pulses and oilseeds under varied agro-climatic conditions. The feed back provided by these demonstrations had helped the scientists to formulate and/or modify research projects to further increase production per unit area per unit time.

Operational Research Project

To address the problem of identifying the operational constraints which impede successful transfer of technologies from the research stations to farmers' field and to extend the concept of National Demonstration on an area or water shed basis, the Operational Research Project (ORP) came into operation in 1975. The aim of ORP was to look at the entire farming system and to identify the socio-cultural and socio-economic constraints as well as technological and administrative constraints with a view to specifying the steps necessary for deriving full benefit from the untapped production reservoir existing in an operation area. The concept of "Social Audit", besides the scientific and financial audits was introduced for the first time under ORP.

Krishi Vigyan Kendras

Closely on the heels of ORP, the Krishi Vigyan Kendras (KVK) were established in 1975 to impart training to those extension workers who are already employed and to the practicing farmers, farm women and rural youth who wish to be self employed. KVKs are the grass root level vocational training institutions designed for bridging the gap between the available technologies at the one end and their application for increased production at the other. The concept of human resource development in agricultural development was given paramount importance for the first time through KVK. Normally, a district is the operational area of KVK.

Lab to Land Programme

It was during 1979, when the Indian Council of Agricultural Research was celebrating its Golden Jubilee, the Lab to Land Programme was thought of. This programme emanated from the fact that hardly 30 per cent of the technologies alone generated from various research institutions reached the farmers and a lot more of them still remain within the four walls of such organisations. Moreover, it was felt ideal if the researchers themselves transfer the developed technologies to the field for gaining first hand knowledge about the complexities of technology transfer and the constraints involved in it. With this concept in mind, various ICAR Institutions and State Agricultural Universities were identified as implementing centers and entrusted with the responsibility of carrying out the Lab to Land Programme in an adopted village amongst the selected number of farm families that was assigned to each of these centers. Under this programme each center has to identify one or two most viable technologies developed by them for transfer to the adopted village. Efforts were also made towards overall development of the selected farm families by mobilizing all the developmental programmes in operation at the adopted village. A multidisciplinary team of scientists constituted at each center was responsible for overall implementation of the programme.

Impact Ananlysis

In many of the studies employed to measure the impact of various agricultural development programmes, it was time and again pointed out that inspite of the production increase due to various programmes since 1950, there was no uniform growth across the country and there were regional imbalances in the technology adoption as well as disparity among the farmers in absorbing the technology. It was further observed that "Green Revolution" rather failed to address the needs of such environments which are complex, diverse, risk prone, operate under low in put and enjoy less govt. support. A close scrutiny of the technologies generated in the research system and their transfer to the farmers' field revealed that they are not very much appropriate to various production systems. This pointed to the inherent weakness in the present technology and their mismatch to such production systems which are dissimilar to research institutions. This understanding necessitated the need for technology assessment and if needed refinement through participatory approach to identify location-specific appropriate technologies. This concept has been crystallized into a new approach in technology transfer called Institution-Village Linkage Programme (IVLP) by ICAR which lay emphasis on problem diagnosis, technology intervention and assessment through users participatory research mode.

Institution-Village Linkage Programme

It is an innovative project initiated by the Indian Council of Agricultural Research (ICAR) on a pilot basis from 1995-96 which was later brought under the World Bank funded National Agricultural Technology Project (NATP) since 1999. It is different from the earlier first line extension efforts of ICAR, in the sense that it lays emphasis on the research aspect through the participation of farmers to be carried out by the multidisciplinary team of scientists. Moreover, IVLP is a production system oriented project with agro-ecosystem analysis of the adopted village as the basis to identify problems, prioritize them and find out technological intervention points which are further developed into action plans to overcome the problems through assessment and refinement of technologies. Active participation of the farmers has to be ensured throughout the implementation of this project and in this regard various PRA tools and techniques are to be employed to study the agro-ecosystem, develop action plans, implement the same and assess their appropriateness to the village.

Concept of IVLP

The IVLP is conceptualized based on the realisation of the fact that majority of the agricultural technologies are developed and perfected at the research institutes under ideal conditions which seldom exists under small and marginal farmers situation, that form the bulk of the farming community in India. The impact analysis of the various transfer of technology (TOT) projects of yester years brought to the lime light that the cause for non adoption/ partial adoption of agricultural technologies by the farmers operating under complex, diverse, risk prone (CDR) production system really did not lay with the farmers or extension system or input supply system as believed generally, rather it was the technology that was found to be a mismatch to the production system of the farmers. It is against this background, the need to evaluate agricultural technologies for their appropriateness to the environmental and socio-economic conditions of small and marginal farmers was felt strongly. Hence, a more holistic approach encompassing the detailed analysis of the agro-ecosystem of an adopted village, problem diagnosis, assessment and refinement of identified technological interventions etc. with farmers participation as the underlining principle was thought of by ICAR and named as Institution-Village Linkage Programme (IVLP) to address the problems of farming community particularly of CDR system.

Objectives of IVLP

The main objectives of the project include

(i) To introduce technological interventions with emphasis on stability and sustainability along

with productivity and profitability taking into account environmental issues in well endowed and small production systems.

(ii) To introduce and integrate appropriate technologies to increase the productivity with marketed surplus in commercial and off farm production systems.

(iii) To monitor socio-economic impact of technological interventions for different production systems and

(iv) To identify extrapolation domains for new technology/technology modules based on environmental characterisation at meso and mega levels.

The various steps envisaged in the implementation of the IVLP are detailed below.

Selection of Operation Site

Village is the basic unit of operation of IVLP project. While selecting a village, the factors such as proximity to the implementing centre, size of the village (800-1000 farm families) contiguity, institutional development, availability of various kinds of production system, co-operative and willing nature of farmers, etc. should be given due consideration. Giving due weightage to the above criteria and taking all the above aspects into consideration, Chenkal village in Neyyattinkara Taluk of Thiruvananthapuram district, Kerala was selected for implementation of IVLP. It is situated about 35kms. south of Thiruvananthapuram city. The Chenkal village (west) has been chosen with 600 families, out of which 570 derive their income from agriculture and allied activities.

Constitution of Multidisciplinary Team

The project is to be implemented by a multidisciplinary Core Team of 4-5 scientists drawn from the host institute with one amongst them acting as the Nodal Officer. It is the primary responsibility of the team members to select the village, prepare a project document, implement the action plans contemplated, monitoring of day-to-day activities and submission of periodical reports. Apart from the Core Team, an Optional Team is also constituted with members from disciplines other than those of Core Team from the host institution as well as from other institutions to render technical guidance as per the need of the project. The core team of CTCRI consisted of 5 members, a plant breeder, an agronomist, a plant protection scientist and two extension specialists, one each acting as a Nodal Officer and Core Team Leader.

Agro-ecosystem Analysis

Agro-ecosystem analysis is a recent concept of multidisciplinary nature and can deal all levels of hierarchy of agro-ecosystems, from field through farm, village and water shed, to region and nation. It also provides a technique of analysis and packages of technology that focus not only on productivity, but also explicitly and on trade-offs between them. The complexity of the system in terms of its dynamic consequences can have four system properties which together describe essential behaviour of agro-ecosystems. These are productivity, stability, sustainability and equitability. To understand the system properties, the pattern analysis is widely used which includes four patterns that reveal the key functional relationship that determine system properties. Three of them, viz. space, time and flow are known to be important in understanding the system properties and are also neutral with respect of scientific disciplines. The fourth pattern-decision reflects the process of human management of agro-ecosystems. The agro-ecosystem analysis was carried out by employing various PRA and RRA tools and techniques such as key informant interview, focused group discussion, village transect, mapping, diagramming, matrix ranking etc.

Pattern analysis

Space

The spatial patterns of the village is understood by village map and village transect. While village map gives a total picture of the village with regard to domicile pattern, land utilisation, availability of various facilities etc., the transect represents the cross section of the village indicating the topography, soil type, crops grown, livestock details, irrigation source, problems etc. of the village.

Time

The time patterns are best expressed by graphical representations. Historical transect, rainfall pattern, cropping pattern, seasonality of pests and diseases, yield trends, labour availability, fodder availability etc. are some of the key parameters in the analysis of time patterns of the village. This in turn will help in identifying the periods in the year when the timing of operations and the availability of resources is critical for productivity and stability. Long term changes in production, prices, climate etc. and the stress and perturbation occurring in the agro-ecosystem are also analyzed to reveal productivity trends and to measure the stability/instability of production.

Flow

The patterns of flow represent the transformations of energy, materials, money, information etc. in the agro-ecosystem. Flow diagrams are used to depict the interrelationship between the various elements. The resource flow diagram of the adopted village revealed that 90 per cent of the cassava produced is going out of the village through sales and only 10 per cent is consumed within the village. Mobility map and livelihood analysis are the other important flow patterns used to get an idea about the information seeking behaviour of the farmers and their allocation of resources towards income and expenditure pattern.

Decisions

Decisions, mean individual farmer's choice in the selection of enterprises, crops and crop varieties as well as management practices, such as what to grow, where to grow, how and from where to purchase inputs etc. which occur at all levels in the hierarchy of agro-ecosystems. They are mainly represented in the form of matrices and they help in identifying the priorities given by the farmers as well as the selection criteria used by them in arriving at such priorities.

Analysis of System Properties

Productivity

The information on productivity of a crop over the years in the village is collected and compared with that of the productivity at block, district and state level. This is then categorized as low, medium or high and the scope for improvement if any is also identified. For example the productivity of cassava (20 t ha^{-1}) in the village, though higher than the productivity of the respective block and district is much below the realizable potential as opined by the experts. This implies that farmers can achieve the potential yield by replacing the existing land races with high yielding varieties. Hence, the cassava productivity of the village may be regarded as medium.

Stability

In this analysis, the consistency of the production of a crop in the village over a time period is assessed. In the adopted village Chenkal, as against many of the crops like paddy, banana, coconut, vegetables etc. which showed variability in stability of production, cassava exhibited consistency in the stability with steady increase in productivity from a meagre 9 t ha^{-1} during 1950 to the present level of 20 t ha^{-1}.

Sustainability

Sustainability refers to the ability of the village to carry forward whatever production increases that had taken place in a crop. In other words, it measures how far the productivity gains are being sustained in the village. Owing to the steady increase in cassava productivity during the past five decades, there is high sustainability of this crop in the adopted village.

Equitability

Equitability is estimated using wealth ranking and it was observed that there is no equal distribution of wealth among the villagers with agricultural labourers accounting for more than 65 per cent of farm families. However, as far as the cassava cultivation is concerned there is high equitability, since it is grown uniformly by all the sections of the farmers cutting across land holding size, resource availability, caste etc.

Problem Diagnosis and Prioritisation

By effectively utilizing the technique of key informant interview, focused group discussion and direct observation, the various problems confronting the production of crops and other enterprises are diagnosed. The basic root causes for problems are also enlisted and grouped into bio-physical and socio-economical. In consultation with the farmers and concerned experts in the field, the causes are prioritized for possible technological interventions. This entire exercise is depicted in the form of problem-cause relationship diagram indicating the different bio-physical and socio-economic causes for a particular problem. Then the point of interventions are identified in order to overcome the problems through possible technological interventions.

Legitimisation of Problems and Solutions

Legitimisation is a process by which problems and solutions are presented and discussed with the villagers for getting their concurrence on these, done through focused group discussion in which farmers, Core Team members and experts take part in this exercise.This forms the basis for developing appropriate action plans to be implemented to overcome the problems identified.

Action Plan and Implementation

Keeping the problem-cause relationship as the basis, action plans are prepared for each technological interventions with the active participation of farmers facilitated by IVLP Core Team. The action plan gives an account of the micro-farming situation, problem and its causes, intervention points, potential solutions, nature of intervention, treatments, critical inputs etc. The technological interventions contemplated under the action plans are implemented in the selected farmers field either as on-farm trial or verification trial. At each and every stage of the trial, the participating farmers are involved in assessing the interventions using various performance indicators such as technical observations, economic observations, farmers reaction etc.

Lessons Learnt

During the course of 4-5 years since the inception of the programme in 1999, CTCRI has implemented over 80 technological interventions under various production systems existing in the adopted village viz. paddy based production system, banana based production system, vegetable based production system, coconut based production system, cassava based production system and mixed farming system. A detailed assessment of these interventions and their pattern of adoption by the farming community explicitly brought out the fact that there is a differential pattern in the adoption based on technological intervention and the socio-economic and cultural milieu under which the farmers operate. Based on the experience gained under IVLP, the technological interventions can be broadly classified into three categories. They are

(i) Interventions requiring no precondition for adoption

(ii) Interventions whose adoption is possible only with certain preconditions and

(iii) Interventions that is effective only at community level.

Interventions Requiring no Precondition for Adoption

The high yielding varieties of crops introduced and assessed to be appropriate to the adopted village under the programme are the classical examples for this kind of intervention. Of the 15 paddy varieties assessed for their performance, the high yielding semi tall varieties "Athira" and "Uma" were highly successful owing to high grain yield coupled with high straw yield, bold grains, high volume of cooked rice, taste, market preference, suitability to both seasons, field tolerant to pests/diseases etc. Similar is the case with regard to tissue culture banana var. Robusta (Grand naine) and Nendran; "Arun" variety of amaranthus (both as pure crop and as intercrop in banana); "Kaumudhi" of snake gourd; "Preethi" of bitter gourd; "Varadha" and "Mahima" of ginger and "Sree Priya" of white yam (Both as intercrops in coconut). In general, the varieties which were found ideal and suited to the socio-economic conditions of the farmers and micro-climatic conditions of the adopted village, do not require any other condition for their dissemination and adoption. Cutting across the barriers of caste, income, land size etc., once the planting materials of these varieties are made available under IVLP, majority of the farmers started adopting them. Presently, the two identified paddy varieties occupy around 80 per cent of paddy area in the village and gradually spreading to the neighbouring villages. The other varietal interventions too have made considerable inroad in the adopted village without any preconditions.

Interventions whose Adoption is Possible Only with Preconditions

Many of the interventions relating to agro-techniques and small scale mechanisation in crop production fall under this category. Especially the nutrient management, chemical weed control, use of drum seeder for sowing, use of rotary weeder etc. in paddy cultivation required certain preconditions for their effective adoption. In case of nutrient management, application of 'P' alone as basal at the time of transplanting was beneficial in increasing the yield; however large scale adoption of this intervention was constrained due to non-availability of P fertilizer, single super phosphate/mossorie phosphate/rock phosphate at the village outlet. At many times, the availability of particular brand of fertilizers that decides the course of nutrient management, rather than the effectiveness. The case of weed management was little different from this. In addition to the availability of the recommended weedicide at the village outlet, its effectiveness at the field level also depended on the regulation of irrigation water during the course of weedicide application. Since, many of the paddy fields have been converted to grow other crops, it is beyond the control of the individual farmer to do any thing with water management. As a result, both the treated as well as the hand weeded plots yielded same under IVLP. While the use of drum seeder needed regulation of water at the time of seeding, the use of rotary weeder was effective only under line sowing which the farmers of the adopted village seldom practice. Hence, considering the micro-climatic environment, necessary modifications/refinements are required in such interventions for large scale adoption by the farmers.

Interventions Effective Only at Community Level

Plant protection interventions aimed at pest/disease control were found to be effective only at community level or on an area basis. Experience under IVLP revealed that many of the farmers go blindly by convention, when it comes to pest/disease control in crops like paddy or vegetable. Once a pest is noticed in a field, spraying is done by that farmer and others follow the suit. Invariably, the plant protection chemicals are used indiscriminately which often leads to residual toxicity and the pests becoming resistant to that chemical. Against this background, it was found extremely difficult to motivate the individual farmers to go in for integrated pest management measures. This became a futile exercise and has not resulted in effective adoption by the individual farmers. Hence, it is deduced that plant protection interventions need to be taken up only at community level or at least on an operational area basis by mobilizing all the farmers of that area and organizing a plant protection campaign; only then it will succeed.

Conclusion

The Institution-Village Linkage Programme is a holistic approach, treating the production system as a single unit and assessing the technological performance to increase the productivity, sustainability and equitability of the system with active participation of the farmers. This approach starting from the problem diagnosis to the final technological assessment was very useful in identifying the most profitable and appropriate interventions under each production system. By this, the time lag between technology generation and technology dissemination has been considerably reduced. Moreover, the spread of the technologies was also found to be faster amongst the beneficiaries, since the farmers themselves select the best technologies from among the plethora of technologies made available under the programme. This IVLP approach needs to be institutionalized and organisations like State Agricultural Universities, ICAR Institutions, Krishi Vigyan Kendras, Research cum Extension wing of input firms etc. can adopt this approach in identifying suitable technologies for their area of operation, particularly at the micro-climatic level. Once such technologies are identified, the regular technology transfer system namely, the development departments operating at various states can take up these technologies for large scale transfer to similar domains.

e-Extension on Transfer of Technology—An Experience

F.R. Sheriff

ABSTRACT

Tamilnadu Veterinary and Animal Sciences University established three Village Information Centres to assess the interest on accessing information through modern IT Tools. The centres are located at Vellore, Tiruchirapalli and Madurai districts of Tamilnadu. The data from July 2001 to July 2004 was collected and summarized as follows:

The Village Information Centres are being maintained by Steering Committee belonging to the same Village Community with monitoring support of Veterinary University Training and Research Center as hub centres. Village community felt happy on establishing the centre in their village. They were also given basic training on computer literacy by the hub centre of the University. The information tools such as computer, printer, internet facilities, television, radio and print materials were supplied to find out the use of different media on accessibility of the required information. The village communities were supported by technical staff of the University for computer literacy. Women found it more useful after the training for themselves as well as for their children for knowledge gaining. Some of the technologies like rabbit farming was taken up for revenue generation after looking through CD. The villagers were also able to get information about disease on sheep and obtained immediate assistance from the Training and Research Center of the University, thus saving the livelihood of sheep farmers. Even some of the illiterates have started using email facility to correspond with their relatives and their kiths and kins. During the year 2001-2003, 27% of the people used electronic media for information access. After two years, the technical staff were withdrawn and allowed village communities to take care of their centres. During the last one year period (2003-2004) the information access through electronic media stands second in priority while print media being the first. Many people have not used the Video media, as most of the villages are covered with cable TV network and this media is being used for entertainment purpose. The main problems faced are frequent power failure, illiteracy, connectivity and break down of equipments in rural places. People are gaining confidence in the usage of information tools for getting information as well as to share the information. The information tools are helping in self learning and also to share the knowledge and hope to see further progress and growth, in future.

Introduction

Tamilnadu Veterinary and Animal Sciences University started an experiment of rural e-extension for transfer of technology in three villages. Our primary aim was to find out whether the rural people who are our real clients can access information on their felt needs electronically. In Tamilnadu, most of the villagers have colour televisions with remote controls and also get local cable connections to view different programmes telecast by different channels. Most of the youngsters hire VCD and CDs and play it at their house for entertainment. In the present situation we thought it is worthwhile to explore the possibility of information access by the rural people on their technological needs. We established three Village Information Centres at Chitteri village in Vellore District, Kuzhumani village in Tiruchirapalli District and Pudhuthamaraipatti in Madurai district of Tamilnadu. These centres were established in July 2001 and till July 2004 the data were collected.

Selection of Villages

We identified Vellore District as it is about 120 km from Chennai, a metropolitan city and adjoining to state of Andhra Pradesh. Tiruchirapalli was selected as it is located on the central part of Tamilnadu. Madurai district was selected as it is on the Southern part of the State. In each of these district our University has a Training and Research Centre. The centre staff were given responsibility of selecting a suitable village. The criteria of selection were

- A remote village with facilities of conveyance
- Facility of electricity and phone connection
- The villages should have main occupation of agriculture/animal husbandry/horticulture
- Willingness of establishing an information centre
- To meet the electricity, telephone charges and provide free space to accommodate the centre
- Village to have 500 to 1000 families only

Based on the above criteria each Training and Research Centre identified 4 to 5 villages. A survey instrument was applied and randomly picked up 10 to 15 men and women in each village to find out the information needs, trust and satisfaction of the information obtained. Their demographic details were also obtained. Based on the interest and the felt needs of the villagers the three villages were selected to establish the village information centres. The Training and Research centres in the district were established as Hub centres under the overall control of the Directorate of Extension Education of the University.

Information Tools and Establishment of Connectivity

The hub centres were provided with a computer, printer, web camera, speakers, telephone and internet connectivity. Each of the Village Information centres were provided with a Computer, printer, web camera, speakers, telephone, internet connectivity, printed books, journals, newspapers, video compact disc player, television, radio with tape and CD Player. For Vellore and Tiruchirapalli village information centres dial up internet connectivity was provided, Madurai Village Information Center had CorDect wireless local loop (WLL) technology for internet connectivity.

Made a Beginning

A steering committee composed of, the local Panchayat leader, Progressive farmer, middle level farmer, small farmer, landless labourer, women, youth, bank official, postal employee, trader, business man, self help group animator and the Hub Center officer as member secretary. In each village a pre-tested questionnaire was administered with 250 family members to assess the demographic details, information needs, and educational level. The steering Committee was given the responsibility of managing the Center by taking appropriate decisions.

Focus group exercise was done to educate the community on their present status and the ways of improving their knowledge. Based on the decision of the steering Committee more than 10 volunteers were given training on basics of computer applications. Audio and Video CDs containing Animal Husbandry Practices, Fishery Products, Value added products, were supplied to the village information centres. Based on the felt needs of the villages a tamil portal was prepared containing District profile, Communication details, Agriculture, Animal Husbandry, Horticulture, Fisheries, Home Science, Education, Government—departments, marketing places and frequently asked questions and hosted in our university website: www.tanuvas.org/rural. The content was prepared by the hub centre staff and the Directorate of Extension Education. The Collaborators for content preparation were College of Agricultural Science, Tiruchirapalli, Tamilnadu Agricultural University, Tamilnadu Women Development Corporation and the District collectorate, More details was collected by the Junior and Senior research fellows posted in the village Information Centers.

Our external Collaborators in the scheme were the staff of M.S. Swaminathan Research Foundation, n-logue communication, Prof Royal D. Colle Department of Communication Cornell University, Ithaca, USA and the funding source is from International Development Research Center, Canada

Social Innovation

After the establishment of Information Center and teaching the basics of computer operation the women, men, youth are being respected by the family members and their morale is boosted. Opening of information center in their village and leaving it to the care and benefit has a marked change in terms of social aspect and self-esteem of the individuals. They feel that their status has improved by possession of the computer center. The same respect is being extended by local government authorities like municipality, and panchayat board for having and accessing computer by the villagers.

The SHG women provided with information communication technology tools are continuously using personal computers for their knowledge building and also teaching other members on its usage, which is likely to be a continuous process. SHG women expressed that they have seen the computer in movies and shops. But they never had an opportunity to use the computer. So they felt proud to have the computer and showed a great amount of interest to learn the computer operations. Being members of the self help groups, the women, in general and resource poor rural women in particular who are below the poverty line can wield greater advantage in seeking technical information, inputs training etc., for their income generation.

One of the SHG members Mrs. Anitha, 31 years old woman asked the action researcher in the beginning What is the use of computer education? If you give skill training programme it will be useful to us. After one year of experience with Computer literacy she said that earlier she had inferiority complex due to her low level of education. Now she feels that she is equal to an educated woman. She was telling that she could operate the computer as an educated woman.

Mrs. Valarmathi animator of Indhu SHG told that her father-in-law was a very strict person and controlled his family members. He would not allow them to mingle with others. He used to close the doors and windows if there was any quarrel on the street. Mrs.Valarmathi had misunderstanding with her brother–in-law's wife as she scolded her a lot. They have cheated them on sharing family property. She decided to improve her status in life. She has started a group and she asked their neighbors to join in the group. Her group is the oldest group in the street. As this being the family background she showed greater interest to learn computer instead of quarrelling with relatives. She has planned to make her group as the best group in this district. They do not know anything about skill training. She stated that "We are managing the group for thirty months and nobody had taken steps to give skill training for us. But the university gave computer training and skill training for us. Now they are preparing six products and the products are getting popularity through the website. She is mentioning here about www.rural Bazaar.tn.nic Tamilnadu Government has developed a website www.ruralbazar.tn.nic.in to market the products prepared by SHG women in Tamilnadu. This group received invitation from DRDA (District rural development agency) officials to participate at SHG products exhibition in Mumbai. The Indian Bank Manager came forward to give loan to carry on business on a larger scale. This is all because of establishment of information center. Now the group has become more active and the hopes of members are increasing.

From their view, empowerment is

- Awareness
- Self confidence
- Boldness
- Decision making
- Economically sound
- Good education
- Freedom

Expectations on Family Future

SHG members are not only interested to learn computer but they also want to give computer education to their children as they are not in a position to send their children to matriculation school. SHG member's children are studying in Government schools. So center is the only access point for them. Even though lots of computer institutions are available in the nearby vicinity, they are not in a position to pay exorbitant fees.

The members are not getting adequate skill based training programme for income generation. Women are not only interested on information but also interested in the hands on training, as without it, the information alone will not lead them anywhere. This has motivated us to provide different skill training programme to the members.

Various CDs are given to the members for self-learning. Mrs. Devi was rearing rabbits in her house. She started rearing rabbit as pet animal. Now she has 20 rabbits in her house. She does not have any idea about proper rearing of rabbits. She kept all rabbits in open terrace of her house. The place was dirty. Once she came to the center with her daughter. She eagerly watched the rabbit rearing CD of Tamilnadu Veterinary and Animal Sciences University. After that she modified the cage and maintained the area without dirt and waste materials. She prepared feed scientifically for her rabbits according to the formula. Now she wants to make it as a successful business. She started selling rabbits to her friends, peers, relatives etc., and started earning money.

Information is available on various subjects but to access it one has to be well versed in English language. Government of Tamilnadu is continuously developing various websites; the required information is yet to be made available. It is time for the educational institutions and the Government authorities to develop information through CD based updates in local language.

Mrs. Kotteeswari is a member of Indus SHG. She studied up to third standard. She expressed that "I used to draw picture in computer. I could not remember everything. My daughter can operate computer better than me"

Mr.V.Gurusamy is a resource poor watchman and aged about 60 years who is living in Puduthamaraipatti village. His son Mr. Narasimhan is doing Ph.D in Niigita University of Pharmacy and applied life sciences, Japan. Mr.V.Gurusamy used to contact his son through letters, which takes one month time to get reply from his son. With the help of Senior Research Fellow of Village Information Centre email id was created to Mr. Gurusamy and subsequently he got trained in internet and sending e-Mail. This made him possible to contact his son and he too finds it convenient to respond over mail. Now without others help he is capable of sending mail to his son. His family members were very happy by getting immediate reply from their son through Village Information Centre.

In Melur taluk of Madurai District, computer kiosk centers were established in 40 villages by Dhan Foundation. All the kiosk centers and the Village Information Centre of Puduthamaraipatti of TANUVAS-IDRC were connected to internet through cordect wireless technology. The kiosk operators of nearby villages used to send veterinary queries to the Village Information Centre of the Puduthamaraipatti since it is managed by two veterinarians. On 9.10.02 the Village Information Centre of Puduthamaraipatti received a mail from Miss. Suganya, kiosk operator of T.Ulagupitchanpatti village stating that six sheep of Mr.Pitchai and two of Mr.Pandi were died due to some disease and around hundred sheep were suffering with diahorrea and other illness. Immediately the Village Information Center staff contacted the Professor and Head, Veterinary University Training and Research Center (Hub centre) and conveyed the message over e-mail and phone. Then the Village Information Centre staff with a specialist (Pathologist) rushed to the spot (T.Ulagupitchanpatti) and inspected the problems of the sheep. Based on the symptoms they came to the conclusion that the sheep were affected by disease called "Enterotoxemia" a bacterial disease. Samples were collected from the affected animals and sent to the lab for confirmative diagnosis. The affected animals were recovered completely. Immediately around 623 (worth Rs.6, 23,000) sheep were vaccinated with Enterotoximia vaccine. All the sheep population of the villagers was saved by timely information services available at T.Ulagupitchanpatti and Puduthamaraipatti.

Usage of information centres at three villages during the scheme period and subsequently are as follows

User Profile of Village Information Centers—Chitheri, Kuzhumani and Puduthamaraipatti

Sl. No.	*Media used*	*July to July 2001*	*2003*	*July to July 2003*	*2004*
		No. of people	*Percentage*	*No. of people*	*Percentage*
1.	Print Media	10541	59.29	5743	52.4
2.	Electronic Media	4883	27.46	1809	18.4
3.	Audio Media	1906	10.72	935	9.5
4.	Video Media	202	1.14	712	7.3
5.	Others	248	1.39	1217	12.4
	Total	17780	100	9816	100

Lessons Learnt

Based upon this experience acquired over the project period, a number of useful suggestions have emerged with regard to advising future applications of the information centre model for community development.

1. The project site should be selected carefully. Nothing can be done successfully with out the target people co-operation.
2. Information, which is provided to them, should be need based.
3. Information provided through video and audio were cost effective.
4. Information retrieval should be cost effective.
5. Project should have enough period of time to monitor and analyze the feed back of the project.

Constraints

Even though the project proceeded smoothly, it did experience several constraints, due to frequent power failure, state of illiteracy of the members, poor Internet connectivity, and the equipment failures.

Last but not the Least

A personal computer with operating system, telephone connectivity and modem may not cost more than Rs 60,000/ per center. Rural information centers can be established in villages, which can open up markets for technology, which is at present invisible. We can create bottom up revolution by giving ICT facility to the women group and resource poor village community. Computers will enable them to learn new skills, which could alleviate poverty and create self-employment opportunities.

16

Participatory Curriculum Development Through Farmer Field School

M. Anantharman and S. Ramanathan

ABSTRACT

Transfer of technology (TOT) mode dominates in the Agricultural Extension System in the developing countries. TOT generally work well with well-endowed farmers and areas and fail to meet its objectives when the technology and the resources of the users of the technology are complex. This situation demands a change in TOT approach which is different from the conventional one and emphasize the people more than the technology. An approach providing a learning situation for the farmers which enhances the farmers to develop their skill through learning by doing on problem diagnosis and experimenting with technologies to overcome the problems is very much essential in dealing with transfer of complex technologies. Farmers field school approach works on these principle of curriculum development by the farmers themselves through their experiences on experimentation with technologies has proved to be successful in transferring Integrated Pest Management/Crop Management technologies among resource poor farmers. This paper details the principles, approaches, models and impacts of Farmer Field School as an alternative model of regular TOT model.

The issue of concern is who makes the choice of technology. Normally those least affected by the choice are the ones responsible for determining the choice, while those who are forced to live with the technology have least say in the matter.

-Hoyzer. N

Introduction

Agricultural Extension can be described as the process of introducing farmers to information and technologies that can improve their production income and welfare (Purcell and Anderson, 1977). The traditional extension system is based on top down approach with emphasis on delivery of pre-designed messages/packages which are usually formulated by research system and delivered to extension agents for packaging and delivery to farmers. In this mode extension agent is an instructor and the farmer is expected to follow the instruction. Thus the aim of agricultural extension is to promote technology packages developed by research institutes to a farmer audience through extension agents. This top down approach ignores existing farmers knowledge, skills and micro production situation of the farmers. Therefore Agricultural Extension in most developing countries has been dominated by the transfer of technology (TOT) mode.

Extension methods usually involve farmer group meetings or demonstration plots, however, small scale resource poor farmers rarely implement the new technology promoted by the agricultural extension messages that they receive. Farmers may be perceived as passive adoptors of technologies but are recognized as the means through which growth in agricultural production can be achieved. It is not difficult to imagine why TOT mode is facing increasing opposition. Farmers often reject or only partially adopt the recommended technologies for several reasons. (1) Technologies are not appropriate under local conditions

(2) Farmers can't afford the investment (3) Farmers do not want to risk switching to a practice with an unknown result (4) Conventional extension messages usually reach only a small number of farmers and those who benefit are usually better off to begin with (Fliert *et al.*, 1996).

When most of the extension services in the TOT mode are struggling to solve the problems of non-adoption of so called 'miracle technologies' other organisations have experimented technology generation and extension models that differ from conventional approach emphasizing the need for the target group participation (Hiemstra *et al* 1992). Farmer Field School shortly referred as FFS is of such approach different from the normal extension approach which pins importance on farmers participatory training and information/technology generation.

Farmer Field School

FFS was experimented and preferred first in Indonesia for training in Integrated Pest Management (IPM). Experiences in several Asian Countries over more than a decade have shown that IPM training and implementation through conventional approaches using preset recommendations and top down extension systems were inadequate (Matterson, *et al,* 1992). IPM, as well as other sustainable agricultural practices, is knowledge-intensive and location specific. Hence it requires the farmers perspectives and convention based on the socio-economic ecological needs. FFS believes not on a set of standard recommendations, but by (1) growing healthy crop (2) field observation (3) conserving natural enemies (4) location specific resources and ecological considerations (5) on-farm experimentation by the farmers (Fliert *et al.,* 1996).

Concept Underlying FFS

Farmers learn better through hands-on experience and when the subject-matter, they are learning is related to their everyday experience and activities. In a field school farmers are encouraged to explore or discover for themselves. Knowledge obtained this way is more easily internalized and put into practice after the training is over. It is a crop season long training with full participation of farmers from decision making to execution. All the sessions take the abilities, knowledge and experience of the participants as their starting point and FFS activities are designed to deepen them (Fliert and Braun, 2000?).

Considering the views on the FFS, it could be defined as a field based learning process within a group in which farmers share experience and try to find their solutions to their problems where field extension worker is a facilitator in developing communication and analytical skills among the farmers for understanding and sharing the principle behind the practices in a training driven research situation. (Ananthajaraman, and Ramanathan 2002). The FFS approach relies on participatory training methods to convey knowledge to field school participants so as to make them into confident experts, self teaching experimenters – and effective trainers of other farmers (Wiebers, 1993).

Training Approach to FFS

The approach to the training in the FFS requires an extension method different from conventional top down farmer training (Rolling, 1994). The comparison of conventional and participatory training approach in FFS is shown in the Table 1.

Principles of the Farmer Field School Approach

Education is the most important that an 'extension' programme can do and farmer is the most important person being educated. Within the educational approach, communication must take place at the field level, dealing with field issues in a dialogue with learners. The field school deals not only with the practice that farmers want to learn about but with farmers as farmers. FFS are conducted for the purpose of helping farmers to master and apply field

Table 15.1: Comparison of Conventional and Participatory Training Approaches

Conventional Training	*Participatory Training*
Training is vehicle for linear transfer of knowledge and skills from experts to farmers	Training is vehicle for joint learning
Training enables participants in "learning to implement"	Training enables participants in "learning to learn"
Training presents "canned" information	Training presents leaner specific information
Training is mainly a teaching task	Training is mainly a learning task
Training is directive and instructive	Training is facilitative
Training expects "absorptive" behaviour by participants	Training expects "sharing" behaviour by participants
Training assumes that participan ts share the same learning style/ environment as trainers	Training assumes that there is a variable learning style/ environment among participants and between them and trainer

management skills. The farmer implements his or her own decisions in his or her own field.

1. **Problem-Posing/Problem-Solving.** Within this form of training problems are seen as challenges, not constraints. Farmer groups are taught numerous analytical methods. Problems are posed to groups in a graduated manner such that trainees can build confidence in their ability to identify and tackle any problem they might encounter in the field.

2. **Principle not Packages.** Educational programs do not promote packages that present weekly atomized

messages. Educational programs take a broad integrated approach to working with farmers based on the principles that farmers need to learn to be better farmers and optimize their incomes. The FFS teaches principles, any activity encompasses several principles, principles bring out cause and effect relationships, principles help farmers discover and learn, principles help farmers learn to learn so that they can continue to learn.

Basic Elements of Farmers Field Schools

1. **Farmers as Experts.** Learning by doing is the learning approach used. Farmers learn by carrying out for themselves the various activities related to the particular farming practice they want to study and learn about. The key thing is that farmers conduct their own field studies. In so doing they become experts on the particular practice they are investigating.

2. **The Field is the Primary Learning Material.** All learning is based in the field. Working in small sub-groups they collect data in the field, analyse the data, make action decisions based on their analyses of the data, and present their decisions to the other farmers in the field school for further discussion, questioning, and refinement.

3. **Extension Workers as Facilitators Not Teachers.** The role of the extension worker is very much that of a facilitator rather than a conventional teacher. The extension worker may take part in the discussion session but as a contributor, rather than leader, in arriving at an agreed consensus on what action needs to be taken at that time.

4. **The Curriculum is Integrated.** The curriculum is integrated to form a holistic approach. Problems confronted in the field are the integrating principle.

5. **Training Follows the Seasonal Cycle.** Training is related to the seasonal cycle of the practice being investigated. For annual crops this would extend from land preparation to harvesting. For tree production and such conservation measures as hedgerows and grass strips training would need to continue over several years for farmers to be able to see for themselves the full range of costs and benefits.

6. **Regular Group Meetings.** Farmers meet at agreed regular intervals. For annual crops such meetings may be every 1 or 2 weeks during the cropping season.

7. **Learning materials are leaner generated**. Farmers generate their own learning materials, from drawing of what they observe, to the field trials themselves. These materials are always consistent with local conditions, are less expensive to develop, are controlled by the learners and thus can be discussed by the learners with others. Learners know the meaning of the materials because they have created the materials.

8. **Group dynamics/team building**. Training includes communication skill building, problem solving, leadership, and discussion methods.

Designing Farmer Field School

The field school is a season long event and has to be programmed effectively to meet the objective of the field school. There can't be rigid programme for field school as it varies accordingly to the types of crop/problem etc. However the procedure suggested below will be useful in designing the programme. (Fliert and Braun, 1999) It has four major parts.

1. Farmer Field School Preparation
2. Farmer Field School Implementation
3. Farmer Field School Evaluation
4. Farmer Field School follow up and scaling up

ICM Farmer Field School Preparation

Preliminary Meetings

Preliminary meetings are desirable to inform the community about the objectives of the FFS, motivate them to participate, and determine together who are the appropriate participants.Ideally, a request for a FFS should come from a farming community. Farmers who request training themselves are usually more motivated and responsible than those appointed by some authority from above. To trigger such requests, field schools can be promoted by conducting a field day where farmers can observe the achievements and process of a previous FFS

A first preliminary meeting is supposed to be of an informative nature to introduce the FFS concepts and raise interest for participation.During the same or a consecutive meeting, FFS participants are selected by the community itself, and the farmers are invited to prioritize the activities proposed for the FFS. They can also discuss problem-solving ideas and compare these with potential solutions originated outside the village. This process is supposed to result in the preparation of a realistic workplan for conducting the FFS in the village. The workplan should specify:

- The date and time of the weekly FFS meeting.
- The location of the field study site.
- The list of FFS participants.
- A weekly schedule of activities for the entire season.
- Designs for field experiments.

The workplan should also contain explicit arrangements for:

- **Provision of a field that will be used as field laboratory by the FFS participants.** The owner of the field should agree to contribute labor for field preparation and inputs such as fertilizer.

- **Financial matters.** If the cost of conducting the FFS are to be borne by the village, it should be clear where the funds will come from for materials, for transportation and possible honorarium of the field school trainer and who will be responsible for providing refreshments. If the costs are to be borne by a national or local program, the details of what will be provided should be specified.

Participant Selection

Experience has shown that 20-25 farmers can constitute a reasonable critical mass in support of FFS in the village.Selection of participants should take place at the first preliminary meeting, attended by all interested farmers from a village. Members should not be selected unilaterally by extension workers or local leaders because this increases the likelihood that some participants are not active farmers or are not motivated to learn about ICM.

ICM Farmer Field School Implementation

ICM Farmer Field School Facilitation

A FFS trainer or facilitator is more than a teacher or an instructor. He plays a complex role as an experienced farmer, questioner, an organizer, and a coordinator. The roles and duties of the facilitator in the FFS are as follows:

- He investigates the main farming problems in the village before starting the FFS, so that he can plan topics to meet participants' needs.
- He arranges for a field to be used for observation and experimentation.
- He prepares all materials required for the special topics and group dynamics exercises before the start of each meeting.

- He observes and analyses the condition of the experimentation plot or the field with the participants, encouraging them to make in-depth observations by asking relevant questions.
- The facilitator should pay close attention to the involvement of all participants, ensuring that no one dominates the discussion, and encouraging silent ones to take part.

Pre-and post-tests

At the first and the last FFS meeting, the participants take a test to evaluate their knowledge level before and after FFS. The pre-test provides the FFS facilitator with some diagnostic information that he can use to adjust the FFS curriculum to the knowledge level of the group. The post-test results are an indicator of progress made during the FFS season.

The facilitator prepares each test by formulating ten questions that relate directly to local field problems. To answer the question, participants choose among three alternatives. When possible, the alternatives should be live sample, for instance leaves with pest damage or nutrient deficiency symptoms, and insect and soil specimens. The pre and post-test questions should be of similar difficulty, and in the local language.

Assignment to Workgroups

Most FFS activities are conducted in small workgroups of five participants, which is considered the optimum number for effective group work and learning. Assignment of people to workgroups can be left to the participants themselves, or arranged through a group dynamics activity held during the first FFS meeting. Changes in the composition of the workgroups should be avoided once the field school is underway, unless a participant drops out early on and can be replaced by someone else.

Experimentation in the Farmers Field School

During the FFS season, the participants conduct experiments in the FFS field. The objective of this activity is to develop experimentation skills and to provide farmers with experience in the evaluation of practices and alternatives.

The participant should be encouraged to set up additional experiments that will give answers to their own specific questions. Alternative experiments are designed by the participants themselves with the help of the FFS facilitator during the preliminary meetings. The selection of research questions for experimentation in the FFS should be based on farmers priorities revealed through a participatory diagnosis.

Experimentation becomes essential to develop management system and practices that suit the specific conditions of the farm and farmer needs to have the ability of experimentation.

Most farmers actually automatically compare their farming methods with those of their neighbours. Some also compare the results of practices between different plots within their farms. Such comparisons yield insights about the range of productivity obtainable on the farm, however they do not permit farmers to judge the relative merits of different management practices. For this farmers need to be able to design, implement and evaluate simple experiments in a systematic fashion. Following are some of the aspects to be taken care of while organizing farmers experimentation.

- Farmers should be familiar with stage of an experiment:
    - ◆ Determining the topic and defining the objective of an experiment.
    - ◆ Design
    - ◆ Preparation

- ◆ Implementation
- ◆ Evaluation

- Understand the basic principles of systematic experimental design.
- Have gained skills to design, implement and evaluate a simple experiment.

Strengths and Weaknesses of Farmers' Experiments

The facilitator reviews the results of the group discussion from the previous FFS session about common problems in the crop cultivation. Ask the participants whether they have ever experimented with the crop. If so, let the participants describe the experiments they have done. The group should analyze and discuss the strengths and weaknesses of those experiments, and draw conclusions about how experiments should be designed.

Rules of the Game

Present the basic principles of systematic experimentation, drawing wherever possible on ideas that emerged from the previous discussion, and making sure to cover each of the points below:

- Prioritizing and determining research topics (varying only one factor at a time in each experiment).
- Setting a clear objective.
- Determining treatments.
- Designing the experiment (replications, randomisation, lay-out, variables to measure).
- Planning the implementation (locations, inputs, labor).
- Implementing the experiment (planting, monitoring, measuring variables, harvesting).

- Evaluating the experiment (simple data processing,- analysis of results, drawing of conclusions).

Planning the FFS Experiments

(a) The participants are divided into four small groups. Each group discusses what they feel are the important topics that they want to investigated on the FFS field. They write their topics on a piece of paper and reach a common conclusion about the ranking of importance of these topics. Each group presents their conclusions to the whole group and then the whole group reaches an agreement about with topic(s) they will choose for the field school experiment (depending on the size of the field they can plan for one or more experiments).

(b) The group defines collectively the research objective and the treatments for each experiment. The title, objective and treatment for each experiments is recorded separately.

(c) In the small groups the participants design the experiments. Each group designs one experiment, and one or more groups may be assigned to work on the same trial. The design for each experiment includes: title, objectives, treatments, number of replications, lay-out in the form of a plot map, materials needed, variables to measure, and processing and evaluation procedures. All groups present their design in plenary session, compare and consolidate them, and come to a final collective design per experiment. The facilitator should watch whether the design fulfill the criteria discussed previously.

(d) Plan the implementation of the experiment together with the participants. When are the

plots planted and treated? Who provides the materials? Who is in charge of monitoring and recording data throughout the season? Make sure everybody knows what is to be done when.

ICM Farmer Field School Evaluation

To identify the strength and weaknesses of each FFS during the season, an evaluation should be conducted by the participants. The evaluation should focus on results, process and impact:

1. Result: What were the results of applying new management practices?
2. Process: How effective were the FFS activities to learn new management practices?
3. Impact: What can participants accomplish by implementing new management practices during FFS in their own fields?

The evaluation is held at a special meeting after the FFS field is harvested. The participants should determine the date, time and place of the meeting.

Evaluation of Results

Evaluation of results is aimed at assessing the effectiveness of technology, particularly for FFS participants who may still be skeptical about new management practice's merit. The evaluation process can also provide inputs for improving the new technology and the FFS learning process. Results are evaluated by weighing the yield of all the experimental plots in the FFS field, analyzing and evaluating the data, and conducting an economic analysis.

During a group discussion, participants compare the yields for each experimental plot and formulate conclusions about the treatments of the experiments and about the technology in comparison with farmer practice. The following question could be raised by the facilitator to stimulate the discussion:

- **How did yield vary with the experimental management practices?**
- **Which practices were associated with a high yield?**
- **What were the production and labor costs (management costs) for each plot?**
- **What was the net income (gross income less management cost) for each plot?**
- **Which management practices were easy to apply? Which were difficult?**
- **How compatible is each practice with the overall farming system practiced by the participants?**

Evaluation of Process

The evaluation of process should assess how well FFS met the needs and expectations of the participants. Criteria for analyzing process include:

- The number of meetings held and number of participants present.
- Reasons for canceling meetings or for absenteeism from meetings.
- Congruence between the special topics covered in the FFS and the local field problems.
- Strength of FFS: What were the most interesting, useful activities?
- Weakness of FFS: what was not interesting, useful? What could be improved? What should be added or deleted? How the curriculum could be improved.
- How did the facilitator perform?

The results of process evaluation provide input to the facilitator for planning subsequent FFS.

Evaluation of Impact

Impact evaluation measures how far the FFS process succeeded in improving farmers' knowledge of the technology and increasing their capacity to apply it.

The evaluation can only measure progress in increasing farmer knowledge and improvement of their skills through a testing process, such as the pre and post-test, and from their own opinions about the success or failure of the field school expressed during the evaluation meeting. Further impact assessment would require follow-up field level observations and /or interviewing of the FFS graduates.

Farmer Field School follow-up

Follow up

After a FFS is over, it is hoped that farmers continue the identified and tested new practices in their own fields, disseminate it to other farmers, and that the cooperation established among them will persist. Before adjourning the field school it is important to discuss what farmers plan to do with their new knowledge and skills. The planning process for FFS follow-up activities should be based on the intentions expressed by graduating farmers, although the facilitator should encourage that attention is given to the following aspects:

- Implementation of the new managent practices in individual fields.
- Group implementation of activities requiring collective action.
- Continuation of (individual and/or collective) experimentation to adopt new practices guidelines to local conditions.
- Farmer-to-farmer dissemination.

To foster dissemination, farmers who have not yet attended a field school can be invited to the FFS harvest. The invitees could include farmers and community leaders both from within the villages and from neighboring villages. At this meeting FFS participants can demonstrate what they have learned during the season to their friends and neighbours. This exposure to the results of the field school will hopefully stimulate new requests for FFS.

Scaling up

Agricultural researches often have too little impact in terms of farmers reached or influence in policy. Scaling up the technology transfer methodology has always been an impediment to development process .The first emphasis on scaling up is on reaching larger numbers of people (Long, 1999). Scaling up leads to more quality benefits to more people over a wider geographic area more quickly, more equitably and more lastingly.

Scaling up approaches could be Project replication, building grass root movement and influencing policy reform. Ulvin and Millar, 2000 categorised the scaling up into four types viz., 1. Functional – a community based programme or a grass roots organisation expands the number and type of activities 2. Political – involving active political intervention 3. Organisation – Improving the organisational strength to improve its effectiveness. 4. Quantitative – increasing membership and geographical area.

Gonsalves (2001) systematized the various perspectives of scaling up as shown in the Fig. 1. There is a spatial dimension wherein technologies spread to larger numbers of farmers over a wider area. The temporal dimension refers to the need to know when a certain technology can be scaled up. The economic dimension refers to the effectiveness of the effort to be kept in mind. The technology dimension often includes the need to diversify the range of technologies or to implement complementary approaches to achieve

Fig. 1. Framework to Various Dimensions of Scaling Up

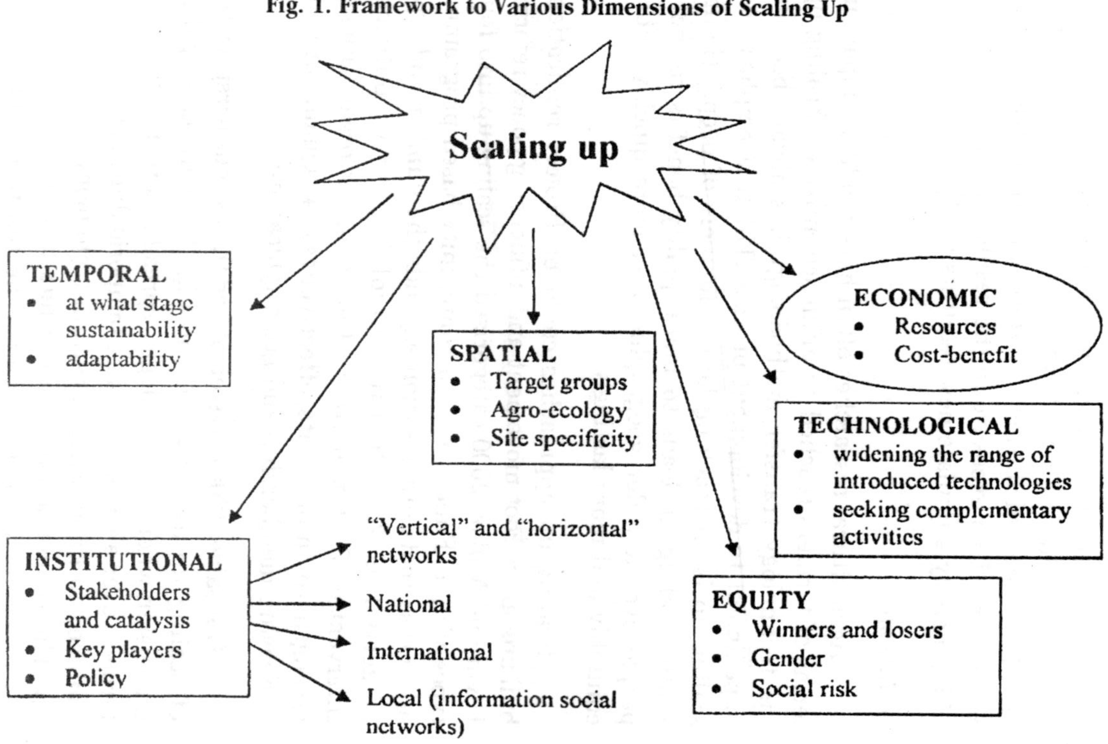

synergism. Scaling up is always a multidimensional, involving technological, process, institutional and policy innovations.

A Basic Sweet Potato ICM Farmer Field School Schedule

A basic schedule for a sweetpotato ICM FFS (model), based on cultivation activities and problems occurring under normal conditions of a crop with growth duration of 4.5 months, is provided below. An adapted schedule, however, needs to be made for each planned ICM FFS, considering specific local conditions. The schedule should be flexible for adjustments throughout the season, when unexpected occur at certain stages of crop development. In addition to the activities and special topics presented in the schedule below, each FFS session should contain the routine activities. (Fliert and Braun, 1999).

Some FFS- Impact Cases

The techniques of IPM have been applied since the 1960's, but many schemes failed to take root with farmers, inspite of demonstrations that natural enemies effectively control many insect pests. IPM training is not effective when simply packaged as part of a top-down extension message. In South and South East Asia, an FAO/Non Government Cooperative programme has pioneered

*Wap**	*Sess-ions*	*Farm calendar*	*FFS activities and special topics (ST)*	*Group dynamics exercises*
-2	0		• Preliminary meeting(s)	
-1	1	Field preparation hoeing, fertilizing, ridging	• Pre-test • Field sanitation • Field preparation and fertilisation • ST-1:Introduction to the sweet potato • ICM Farmer Field School • ST-2: A healthy soil	Line up Family members

*Wap**	*Sessions*	*Farm calendar*	*FFS activities and special topics (ST)*	*Group dynamics exercises*
0	2	Preparation of cuttings	• Preparation of cuttin gs and planting • ST-3: Experimental methodology • ST-4: Healthy seed	How many squares
1	3		• ST-5: Observing the environment • ST-6: Economic analysis of the sweet potato enterprise	Family reunion Nine dots
2	4		• ST-7: A healthy crop sheds skin	The snake
3	5		• ST-5:Observing the crop and its environment • ST-8:Natural enemies: the farmers' friends	Know yourself
4	6		• ST-9: Sweetpotato pests • ST-10: Defoliation experiment	Knotty problem
5	7		• ST-11: Sweetpotato diseases • ST-23: Sweetpotato storage	Follow me
6	8	Soil prep aration for fertilizer	• Soil preparation, fertilizer application • ST-12: Weeds: friends or foes	Trust each other
7	9		• ST-13:Aphids and other tiny insects • ST-14: Pesticides: medicine or poison?	Drawing together
8	10	Apply fertilizer and reform ridges	• ST-15: Fertilisation	Play the rope

*Wap**	*Sess-ions*	*Farm calendar*	*FFS activities and special topics (ST)*	*Group dynamics exercises*
10	11		• ST-16:Vine lifting • ST-17: Field area measurement	Mirror game
12	12		• ST-18:Sweet potato stemborer Defoliation experiment	Drawing a house
14	13		• ST-19:Sweetpotato weevil the blind	Guide
16	14		• ST-20: Cropping pattern item • ST-21: Variety selection	Collectors
18	15		• ST-22: Harvesting and marketing • ST-7: A healthy crop (analysis)	Wayward whispers
20	16	Harvesting	• ST-22: Harvesting and marketing (yield assessment ontest)	
21	17		• Field sanitation • ST-24: Sweetpotato utilisation • ST-25: Evaluation of the sweetpotato ICM FFS • Closing ceremony	

*Wap = Weeks after planting

methods of FFS training rice farmers in the field to understand the plant—pest ecology. (PAN, 1994)

FFS on IPM in Cotton organsed in Garag, Dharwad district of Karnataka empowered women participants to have a good understanding of the Cotton ecosystem, reduced the application of pesticides and 90% male members of their families learned the concept of IPM. (FAO, 2002)

James, (1998) reported on the whole Cassava FFS were well received by farmers and participating extension staff for IPM information testing and transfer, skill development and general human resources development.

Okoth *et al.*, (2002) has reported that FFS on Integrated Pest and Production Management on Cassava and other crops in East African countries Kenya, Uganda and Tanzania were successful in empowering the farmers with knowledge and IPM apart from increasing the competency of extensive systems of these countries. Further results of case studies have shown that the groups are able to recover the whole sum of grant from their commercial plots.

Soil management practices by a multidisciplinary team of Kenya Agricultural Research Institute were validated and disseminated through FFS. In West Aftica, Cassava plant protection technologies were effectively scaled up and disseminated through FFS. (Asiabaka and James, 1999).

References

Ananthraman, M and Ramanathan, S. 2002. Farm Field Schools: Critical Interventions on problem based learning Experiences for the farmers. Paper presented at *International Conference on Lifelong Learning for Social Development,* 13-15, August, 2002. at Thiruvananthapuram.

Asiabaka, C.C. and James, B.B. 1999.Farmer Field School for Particiaptory IPM Technology Development in West Africa. In: Renard,G. et al (eds.) *Farmers and Scientists in a changing environment: Assessing Research in West Africa,* Weikerhein, Germany, Margret Verlag.

FAO. 2002. I have been collecting lady birds-An India Woman demonstrates the impact of small farmer training. *Cotton IPM Newsletters*—December p. 4.

Fliert, E. Van De., Asmunati,R. Wiyonto., Widodo,Y.and Braun, A.R. 1996. From Basic approach to tailored curriculam: Participatory development of a Farmer field School model for Sweet potato. In: UPWARD 1996 . *Into Action Research: Partnerships in Asian Rootcrop Research and Development,* UPWARD, Laguna, Philippines.

Fliert, E. Van De. and Braun, A.R, 1999. *Farmers Field School for Integrated Crop Management of Sweet Potato . Field Guides and technical manual.* Indonesia.

James, Braima D. 1998. Cassava IPM information and its transfer in West Africa. *Integrated Pest Mangement Communications Workshop: Eastern/ Southern Africa,* March 1-6, 1998. Nairobi, Kenya.

Long, C. 1999. Participation in Development: the Way forward. Institute for Development Research, Boston

Matterson, P.C., Gallagher, K.D. and Kenmore, P.E. 1992. Extension of Integrated Pest Management for Plant hoppers in Asian Irrigated Rice. In: Denno, R.F. and Perfect, T. J. (Eds.) *Ecology and management of Plant hoppers*. London: Chapman and Hill.

Okoth, James. R., Khisha, S. Godrick and Julianus, Thomas .2002. The Journey towards Self-financed Farmer Field Schools in East Afr4ica. *International Learning workshop on Farmer Field School: Emerging issues and challenges* 21-25, October, 2002.

PAN, 1994, Farmer first Field Schools are a key to IPM success *Pesticide News* 24: 12-13).

Purecell, D.L. and Anderson, J.R. 1997. *Agricultural Extension and Research: Achievements and Problems in National Systems,* OED, The World Bank, Washington, D.C.

Rolling, N. 1994. Facilitating Sustainable Agriculture: Turning Policy Models Upside Down. In: Scoones, I and Thomson, J. (Eds) *Beyond Farmer First: Rural People's Knowledge, Agricultural Research and Extension Practice*. London IT Publications.

Ulvin, N.P., and Miller, D. 2000. Scaling up: thinking through issues. IIRR Workshop.

Weibers, Uwe-Corrten. 1993. Integrated Pest Management and Pesticide Regulation in Developing Asia. World Bank Technical Paper Number 211. Asia Technical Department Series. The World Bank. Washington, D.C.